Praise for *A Framework for Applying Analytics in Healthcare*

"Dwight McNeill draws an elegant map to strengthen healthcare composed of insights drawn with wit and imagination from four different sectors. The diverse roles of the consumer, customer, member, and constituent in each sector offer new approaches to reforming healthcare that better meets the needs of patients and families."

—**Jessie Gruman**, President and Founder, Center for Advancing Health

"Harnessing the power of health information technology can help transform our healthcare system into the high-quality, patient-centered, and efficient system we need. Dwight shows how healthcare decision makers can learn from the best practices of other industries and use the power of electronic health records and data analytics to improve care and quality."

—**U.S. Senator Sheldon Whitehouse**, State of Rhode Island

"Sharing analytical ideas within companies is commonplace, but McNeill's brainstorm is that healthcare analytics can benefit greatly from adopting ideas from other industries. If you're in healthcare, you'll find that seemingly intractable challenges have already been solved elsewhere. This book will open your eyes to a new set of possibilities."

—**Thomas H. Davenport**, Visiting Professor, Harvard Business School

"Learning from outside your field holds an untapped potential for dramatically improving your performance. This book provides that knowledge and is practical, insightful, and a must-read for the healthcare leaders seeking to transform their organizations."

—**Maulik Joshi**, Ph.D., President, Health Research and Educational Trust

"One of the hottest topics in business education today is how to harness big data for big business results. This book takes the reader on a journey deep into five industries and provides insights and recommendations on how to leverage analytics in very unique and productive ways."

—**William O'Neill**, Dean of the Sawyer School of Management, Suffolk University

"The healthcare industry has a lot to learn from paying attention to the work of other industries. Dwight McNeill has given us a road map for this critical journey. His insights are fresh and compelling. His notion of a Chief Analytic Officer (CAO) is timely. Analytic power will be the fuel for the future, and Dwight's framework provides the platform for a ready understanding of the details. Kudos to McNeill for this far-reaching work. I hope it is not too late for our industry to benefit!"

—**David B. Nash**, MD, MBA, Founding Dean, Jefferson School of Population Health, Thomas Jefferson University

"Dwight McNeill has emerged as an important voice pushing the U.S. healthcare industry to apply innovations learned in other domains to solve some of our persistent operational challenges. In particular, the value and applications of innovative analytics have proven transformative in areas such as finance and retail. Dwight has consistently promoted the notion that healthcare can and must similarly embrace these opportunities in order to thrive through the current period of reform and change. We ignore his voice at our own peril as analytics may well prove to be a key differentiator between healthcare's future winners and losers."

—**Tariq Abu-Jaber**, Vice President of Medical Informatics, Harvard Pilgrim Health Care

"Dwight McNeill not only makes the cogent proposition that the path toward a more efficient, effective, and equitable healthcare system is achievable, but he offers tangible practices, tools, and processes that are immediately applicable to every healthcare decision maker. *A Framework for Applying Analytics in Healthcare* is skillfully researched and makes recommendations that are both innovative and logical. Dwight's comparison and adaptation of analytic 'sweet spots' from the banking, retail, political, and sports sectors are incredibly relevant to today's fast-paced, data-driven information culture. The Health Improvement Capabilities Scorecard (HICS) has the potential to revolutionize how we can turn our current, unsustainable 'sick care' system into a 'healthcare' system, where everyone from hospitals, health plans, physicians, governments, and, most importantly, the patient will benefit."

—**Mark Marino**, Executive Director, Health Leads Boston

"Dwight's model of adopting analytic strengths from diverse industries to transform healthcare holds infinite possibilities for innovation in every industry. It is the foundation for harnessing today's big data and robust analytic platforms to develop tomorrow's products and services that will benefit consumers while growing and transforming business."

—**Marcia Tal**, former Executive Vice President, Citigroup, and Founder of Tal Solutions

"In this book, Dwight has achieved what others only talk about—he's translated towering analytics successes from other industries to healthcare. In an engaging way, he walks through the current challenges in healthcare analytics and is careful not to lose you in the jargon. Next, he deftly highlights the analytics challenges of four industries, highlighting the analogs to healthcare. He ends by synthesizing these observations into adaptations for healthcare and a framework for deploying analytics innovations. This book will help us to stop just talking about applying best practices from other industries and actually do it."

—**David Wiggin**, Program Director, Healthcare and Life Sciences, Teradata Corporation

A Framework for Applying Analytics in Healthcare

What Can Be Learned from the Best Practices in Retail, Banking, Politics, and Sports

Dwight McNeill

Vice President, Publisher: Tim Moore
Associate Publisher and Director of Marketing: Amy Neidlinger
Executive Editor: Jeanne Levine
Operations Specialist: Jodi Kemper
Marketing Manager: Megan Graue
Cover Designer: Alan Clements
Managing Editor: Kristy Hart
Senior Project Editor: Betsy Gratner
Copy Editor: Cheri Clark
Proofreader: Paula Lowell
Senior Indexer: Cheryl Lenser
Compositor: Nonie Ratcliff
Manufacturing Buyer: Dan Uhrig

Published by Pearson Education, Inc.
Publishing as FT Press
Upper Saddle River, New Jersey 07458

FT Press offers excellent discounts on this book when ordered in quantity for bulk purchases or special sales. For more information, please contact U.S. Corporate and Government Sales, 1-800-382-3419, corpsales@pearsontechgroup.com. For sales outside the U.S., please contact International Sales at international@pearsoned.com.

Printed in the United States of America

First Printing June 2013

ISBN-10: 0-13-335374-5
ISBN-13: 978-0-13-335374-7

Pearson Education LTD.
Pearson Education Australia PTY, Limited.
Pearson Education Singapore, Pte. Ltd.
Pearson Education Asia, Ltd.
Pearson Education Canada, Ltd.
Pearson Educación de Mexico, S.A. de C.V.
Pearson Education—Japan
Pearson Education Malaysia, Pte. Ltd.

Library of Congress Control Number: 2013936150

To Joni and Jessie

Contents

About the Author

Dwight McNeill, Ph.D., MPH, President of WayPoint Health Analytics, consults with organizations on healthcare analytics to improve business decision-making, engage customers/patients to produce health, and optimize healthcare financing and delivery. He is a Lecturer at Suffolk University Sawyer School of Management and Lead Faculty Member at the International Institute for Analytics' Health Care & Life Sciences Analytics Research Council. He has been a guest lecturer at several universities, including Brandeis, Harvard, and Duke.

He has worked in corporate settings—most recently for IBM as Global Leader for Business Analytics and Optimization for the Healthcare Industry—and in government settings at the federal (Agency for Healthcare Research and Quality) and state (Commonwealth of MA) levels.

McNeill is the coauthor of *The Value of Building Sustainable Health Care Systems: Capturing the Benefits of Health Plan Transformation* and *Building Organizational Capacity: A Cornerstone of Health System Reform.*

1

Overview

This book is about discovering ideas in faraway places and building bridges to welcome them home.

Introduction

Ideas matter. They are the seeds from which innovations grow. Innovating to achieve competitive differentiation is the top business challenge today. And analytics is the high-octane fuel to power innovations to achieve business breakthroughs.

The premise of this book is that healthcare can learn a lot from other industries. Industries develop deep and unique strengths in certain areas and get very proficient in the associated analytics. Other industries that do not experience the same set of forces do not develop these analytics capabilities. They are blinded from them and their potential performance is constrained. This book provides a guided tour of industries that are somewhat mysterious to healthcare, including retail, banking, politics, and sports. The express purpose is to harvest some ideas and build a bridge to adapt them in healthcare. Indeed, the best analytics from these industries provide insights to address some of the most intractable challenges in healthcare.

The book is divided into three parts. The first part is the discovery process to harvest analytics sweet spots from the four industries. The second part is the distillation and interpretation of the sweet spots into the most compelling healthcare analytics adaptations. The third part provides a roadmap for putting ideas into action and a model to evaluate the adoptability of the healthcare adaptations.

Analytics Innovations

Analytics in healthcare is a paradox. On the one hand, healthcare is immersed in analytics. It is far ahead of other industries in using science, for example, to understand diseases and develop new cures and treatments. But the translation of the science into practice has been disappointingly incomplete. U.S. healthcare faces major challenges, not the least of which are its low standing in the world on key health outcomes, efficiency, cost, disparities, and affordability. And research shows that the odds of getting the right medical service are just a bit higher than 50/50.[1] In a word, the science has just not been put into clinical or business practice sufficiently.

The fact is that we know how to address these challenges, and analytics can be a tremendous support; but there's a blockage. This is due to a number of factors, including the seemingly overarching prerequisite to digitize the business before anything else can be harvested from analytics. This forestalls other forms of analytics that can lead to real benefits for business today. But it is more complicated than that. There is an amalgam of contributing barriers, including the lack of good coordination among actors in the ecosystem, a perverse payment system that does not reward value, complex products and services, ambiguity about who the customer is, professional autonomy, convoluted market dynamics, multiple vested and powerful interests, and a pervasive, risk-averse culture.

In spite of these challenges, the field must innovate to make change happen. Innovation is critical to the success of any business. A recent MIT survey of 3,000 executives across many industries and countries found that the top business challenge is "innovating to achieve competitive differentiation."[2] This was far ahead of the usual business challenges to grow revenue and reduce costs. Innovation rises to the top at this time because the Great Recession required businesses to address all possibilities for cost reduction, and many have developed very lean organizations. Businesses recognize that they need to grow the top line, and not just by extending the old line. And to grow the top line, they have to transform the business.

This is especially true in healthcare. For example, health insurers face commodity prices for premiums, the demand for transparency, and a flip of the business model from business-to-business to business-to-consumer, among other pressures related to healthcare reform. Similarly, providers are facing market and government pressures to improve outcomes, lower waste, and change the underlying revenue model from fee-for-service to global payments. In response, the leading companies are dramatically changing their identities. For example, some health insurers are becoming health companies, and others are viewing claims processing as one of many product lines as they become information companies. The industry archetypes are eroding and innovation is charting a new path.

Innovation is tricky. It is a necessary but a high-risk/high-reward strategy. Companies that do not take the risk can set themselves up for a death spiral. And those that do take the risk need to overcome many challenges, not the least of which is the complexity and unpredictability of implementing innovations. Few innovations succeed. Even with good execution, the innovation might be too early or too late relative to market demand and competitor actions. Or it might prove to be too weak to produce a worthwhile benefit relative to the cost. But for those that do succeed, it can be a game changer and propel the company to new heights.

Innovation starts with creativity. Creativity, whether in art or science, involves looking at things in new and different ways that result in fresh ideas that others regard as beautiful or useful or both. It involves breaking out of established patterns and connecting and rearranging current knowledge. Steve Jobs said that "creativity is just connecting things" like one's experience and knowledge and then "synthesizing new things."[3] There are many methods for generating new ideas, including taking on a different role, for example, of a patient in healthcare, combining ideas from different disciplines, or challenging current paradigms. And, of course, businesses in one industry can learn from businesses in another. Paul Plsek, an expert on complexity and "directed creativity," says that "when we open up our thinking to receive 'fresh knowledge' from outside the healthcare system about customer flow, we may notice the concepts of drive-through windows in the fast food industry, or Internet-based services in banking. This fresh knowledge might feed a creative generation process that results in ideas about how drive-through windows and Internet sites might be used to serve patients in a new primary care clinic."[4]

Ideas begin a process that can ultimately lead to innovations that can make a positive impact on people's lives and the bottom line. Success depends on the wisdom of the idea, the translation of the idea into a design and related strategy to carry it out, excellence in implementing it, and eventually reinventing it.

Ideas can come from comfortable places and unfamiliar places. This book concentrates on the latter, not because they are better, but because they are often ignored. They are ignored because they do not necessarily fit our beliefs about how the world works. We tend to seek out information that confirms our positions and ignore the rest, what is referred to as confirmation bias. So outside-in thinking has an inherent hurdle at the outset. But ignoring outside-in thinking can lead to blind spots. For example, there are legions of dedicated professionals that concentrate on the problems within healthcare with the intent to improve them. The blind spot of this process might be characterized this way: "Removing the faults in a stage-coach may produce a perfect stage-coach, but it is unlikely to produce the first motor car."[5]

When the healthcare industry has looked to other industries for ideas and solutions and subsequently adapted these to healthcare, it has often produced impressive results. For example, quality improvement leaders like Don Berwick from the Institute for Healthcare Improvement and Brent James from Intermountain Healthcare have provoked and worked relentlessly to inspire and educate the field to use process control methods from other industries including the teachings of Deming and methods like six-sigma. And they have demonstrated significant improvements. The frustration has been the slow spread of the innovations that have been proven to be so effective.

Another need for creativity, derived from connecting knowledge from other industries, is to prevent a "Kodak moment." Perhaps you remember the old Kodak moments. These were absolutely beautiful photographs derived from Kodak film products that were published in magazines like *LIFE* decades ago. But the new Kodak moment of 2012 is a not-so-pretty picture of the company filing for bankruptcy. The bankruptcy was due in part to the disruptive technologies of the digital camera and later of smartphones that upended the company's value proposition about its film products. Both digital technologies took good pictures, virtually eliminated cost, and ended the inconvenience of waiting for the pictures to be developed. The other reason for the bankruptcy was that Kodak thought its products were perfect and did not see or subsequently adapt to the threat. However, a competitor in the film business, Fuji, did adapt and thrived. It rethought the use of the chemicals used in the film production business and discovered that they were rich in antioxidants, which are the mainstay of facial makeup products, and it went into the business. There are three issues pertinent to creativity: (1) the need for organizational receptivity to new ideas, (2) the ever-present danger of game-changing, new technology that can make an existing product or business obsolete, and (3) the need to adapt through innovation.

Are there Kodak moments looming for healthcare, and might these be knowable through the sensing of developments in other industries? Of course! And especially in the area of analytics. These are described throughout the book, but here is a glimpse of what's to come:

- Democratization of data, which results in the availability of new data from public "open government" databases and from private "data snatchers"
- Change in the locus of power of data due to mobile technologies and social media
- Deep reliance on predictive modeling to address behavior change
- Allegiance to the customer

Each of these insights could lead to ideas that might flower into innovations, which in turn could result in breakthroughs for business success.

Discovery of Healthcare Adaptations

The premise of the book is that healthcare can and should learn from the best analytics of other industries. Certain ideas will lead to adaptable innovations in healthcare. The process for discovering ideas from other industries used in the book is depicted in Figure 1.1. Note that this approach is based on the firmly held belief that analytics succeeds when it responds to the needs of the business and not when analytics answers—that is, technologies and methodologies—are in search of business questions.

Industry Challenges

The first stage in the discovery process is industry challenges. This stage seeks to understand the industry, goals, context, challenges, and drivers. These ingredients mix together in a crucible that determines how an industry needs to perform to succeed by developing and honing areas of strength. For example, the banking industry suffered a disastrous plunge in trust and revenues due to its irresponsible subprime mortgage lending practices. This not only contributed to the Great Recession, which cut demand for its products, but also led to onerous regulations and oversight that constrained the business, especially in the area of lending. It subsequently developed greater strengths in risk assessment.

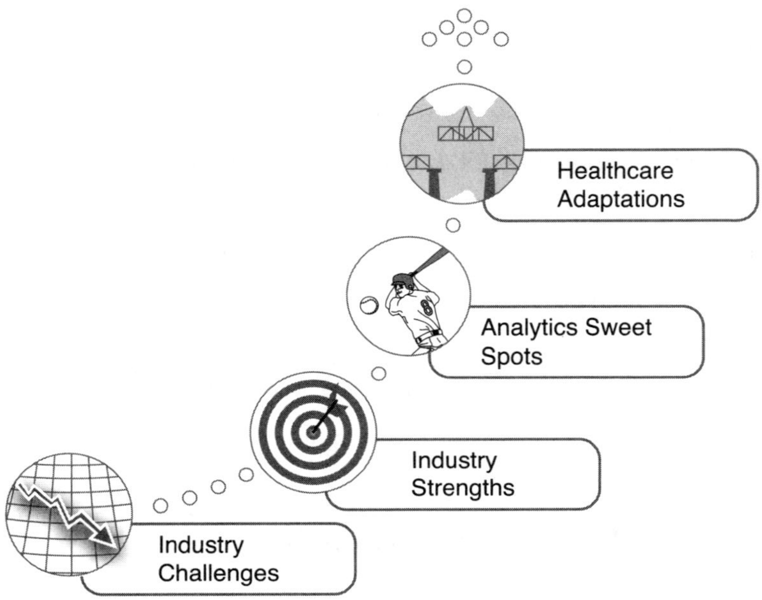

Figure 1.1 Healthcare adaptations discovery process.

The industry challenges are unique to each industry and are detailed in the chapters to follow. There are challenges common to all the industries reviewed in the book, and these are included in Figure 1.2. Note that these industry challenges do not include the generic business challenges, such as increasing revenues, reducing costs, improving the balance sheet, paying attention to customers, and so forth. Rather, these challenges reflect specific game-changing events and trends that portend new waves of threats and opportunities.

The first challenge was the Great Recession. This was not a typical, cyclical recession. In fact, it was the worst economic period since the Great Depression. It shattered customer purchasing power and the demand for products and services.

The second challenge is hyper-competition. This is partly related to the Great Recession and the need to capture dwindling purchasing dollars. But it is also an endpoint of the convergence of the cross-industry challenges described later. Hyper-competition makes demands on analytics to gather and analyze more diverse and larger volumes of data to know the customer more fully. This is most evident in the retail industry, where personal data can help predict what a customer wants to buy before the purchase. Hyper-competition is also evident in political campaigns. The overarching need to find and sway a small number of voters who could spell victory or defeat required a new winning strategy in the 2012 U.S. Presidential election. The winning strategy was to be extremely data driven.

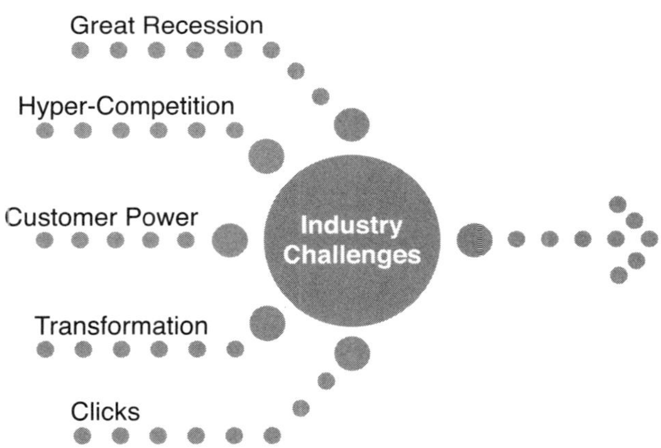

Figure 1.2 Cross-industry challenges.

The third challenge is customer power. Customers are taking the upper hand in their relationships with business because they have the data at their fingertips, in their smartphones and tablets, and they use it to make more informed purchasing decisions. The whole process of buying has been accelerated, yet made more deliberate, because of the availability of price information from multiple sellers on the Internet. Similarly, customers want to do business with companies "their way," on their preferred devices, with access to services 24/7, and expect the business to provide flawless service across channels. Finally, customers want information that makes sense to them and on their terms, which is often obtained from peers and not from marketing departments or government agencies.

The fourth challenge is the need for transformation. There is no sitting still and the race continues to have no finish line. But the difference is that the very nature of the business has to evolve and the pace of change is accelerating. The tried-and-true ways of succeeding in banking, retail, healthcare, and even sports are up for grabs. The need for transformation creates the appetite for innovation.

The final challenge is clicks. Clicks are the sounds of doing business on computing devices. Clicks are challenging the bricks. What could be more indicative of shifting paradigms than the collapse of the structures in which people do business (bricks)? Mobile "rules" for now because it is seemingly a new organ of the body that offers new functionality, is integrated perfectly, and is appreciated. It is almost like a seventh sense for humans. Mobile will give way to the next generation of new (artificial) body parts for sensing and communicating, perhaps as Google Glass eyewear. But the point has been made. Digital devices that provide more human functionality in ways that are "natural" and improve people's capability to sense are game changers.

Industry Strengths

The second stage of the discovery process is the understanding of industry strengths. All industries have unique strengths that, if exploited, can drive business breakthroughs to beat the competition. The unique industry strengths identified in the discovery process and distilled to their essence are summarized in Figure 1.3 and described in the following list.

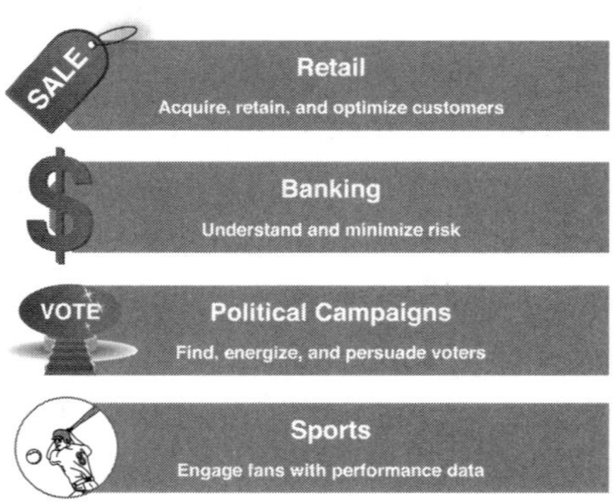

Figure 1.3 Unique industry strengths.

- For retail, it is clear that the overarching strength of the industry is to acquire, retain, and optimize customers. Marketing's customer analytics are woven into this strength.
- For banking, the unique industry strength is understanding and minimizing risk. As mentioned earlier, the industry needed to clean up its lending practices for its very survival. Risk assessment has always been a core element of the industry, but it needed to get a lot better to reduce loan defaults without crimping the volume of loans. This became an industry imperative that necessitated the refinement of the industry strength.
- The industry strength of political campaigns is the laser focus on finding, energizing, and persuading voters. This strength was made more powerful through data-driven innovations.
- Finally, sports has a very unique strength in its ability to engage fans through the full transparency of detailed performance data on its athletes and teams. Athletes are acquired, fired, and improved based on the data. The paradox for healthcare is that sports fans know much more about the athletes who entertain them than patients do about the doctors who make life-and-death decisions about them.

Analytics Sweet Spots

The third stage of the discovery process is identifying the analytics sweet spots that correspond to the industry strengths. The concept of a sweet spot comes from sports. It is the place where a combination of factors results in the maximum impact achieved relative to a given amount of effort. In baseball, for example, it is that place on the bat that produces the most powerful hit. The analogue in analytics is a solution that provides the most power to make the industry strength as successful as possible. In the case of banking, the analytic sweet spot, matched to the industry strength, is a refined FICO score that assesses the capability of a borrower to fulfill promises to repay a loan.

What we have learned is that each industry pushes the envelope in its use of specific analytics, not necessarily because it has more technical sophistication, but because the specific goals, purposes, pressures, and culture of that industry are unique and require better flowering of certain analytic tools. These are the sweet spots that we want to translate and adapt for healthcare.

There are cross-industry analytics themes that shape the manifestation of advanced analytics generally and have a strong influence on the analytics sweet spots for each industry, as highlighted in Figure 1.4.

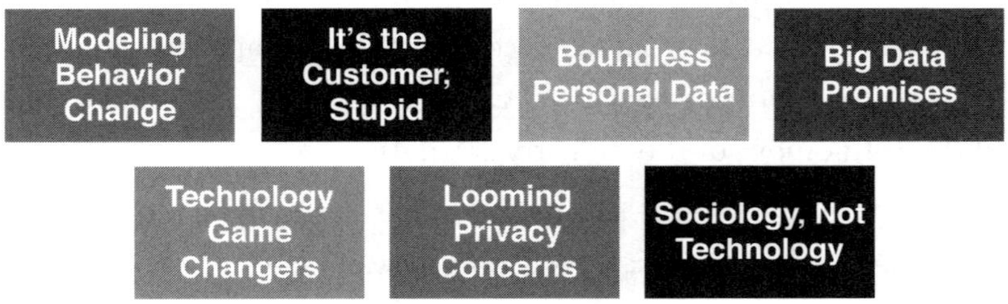

Figure 1.4 Cross-industry analytics themes.

Modeling Behavior Change

Predictive modeling to change behavior is a powerful, advanced analytics method used across the industries. Predictive modeling is commonly used for general business purposes, including forecasting, simulation, root cause analysis, and data mining. Most of the breakthrough applications of predictive modeling across industries in this book focus on understanding and changing behaviors of customers. Examples include the following:

- **Retail:** Determine the probability that a woman is pregnant and her estimated delivery date.
- **Banking:** Determine the likelihood of divorce as a major predictor of loan default.
- **Political campaigns:** Determine the messages most likely to convert an undecided voter.
- **Sports:** Determine what attributes of players contribute to team wins.

It's the Customer, Stupid

It goes without saying that customers are what make businesses thrive or die. Businesses can be distracted and focus on other priorities but do so at their peril.

The industries in this book espouse that the pathway to growth is realized by understanding customers and responding in ways to earn their business. The analytics combo of predictive modeling with "boundless personal data" allows unprecedented views into what makes customers tick.

Boundless Personal Data

We live in a surveillance society. There is a huge business and government appetite to know everything about us. For example:

- **Retail:** "If you use a credit card or a coupon, or fill out a survey, or mail in a refund, or call the customer help line, or open an e-mail we've sent you or visit our Web site, we'll record it and link it to your Guest ID...and we can buy data about you...(such as) charitable giving and cars you own."[6]
- **Politics:** "The campaigns spent over $13 million on acquisition of data like whether voters may have visited pornography Web sites, have homes in foreclosure, are more prone to drink Michelob Ultra than Corona or have gay friends or enjoy expensive vacations."[7]
- **Military:** The National Counterterrorism Center can use any information the government has collected on you, including "flight records, lists of Americans hosting foreign-exchange students, financial records of people seeking federally backed mortgages, health records of patients at veterans' hospitals... this obscure agency has permission to study [any database] for patterns...."[8]

Most of the industries reviewed are using boundless personal data to feed their customer analytics engines. The new data usually come from outside the industry including "open" public databases, data "snatchers," and Web click trackers. A growing industry of data aggregators sell the data and solutions related to it. The large increase in the diversity and volume of personal data, in combination with other analytics methods such as predictive modeling and technology game changers, has been a significant factor in solving business problems and demonstrating the value of analytics across industries.

Big Data Promises

The promise of big data is great and alluring. McKinsey & Company proclaims that "it will become the basis of competition, underpinning new waves of productivity growth, innovation, and consumer surplus."[9] Boundless personal data, as described previously, is a piece of it. But big data goes beyond that and looks to extract meaning from every digital signal that is emitted. It is very likely that harnessing big data will lead to a new world neural system that can measure almost anything. For example, we are told that the technology is nearly there to monitor the heartbeat of every living human being on the planet. But it appears that the technology "hows" are ahead of the business "whys" and "whats." It is unknown how this revolution, like the Internet revolution before it, is going to play out and address business needs. But it is clear that the big promises will take a long time to keep. Lots of research needs to get done, and many hypotheses need to be developed, in order for big data to produce big business results.

Technology Game Changers

Technology advances facilitate the use of data for analytics. Three game changers have been influential:

1. **Clicks:** The Internet has transformed the way businesses communicate, market, do commerce with customers, and collect data about them. One example is the ability to do randomized trials, or A/B testing, of alternative Web site features—for example, how to get the most contributions during a political campaign—on large samples and virtually instantaneously.
2. **Mobile:** Earlier we described the challenges and opportunities of mobile. It is seen as the preferred platform for customer communications across industries.
3. **Hadoop:** Hadoop is an open source software framework that allows for the distributed processing of large data sets across clusters of computers using a simple programming model. It is less expensive and easier to get up and running than commercial applications; for example, it is cloud based and works across hardware. It has greatly facilitated the analysis of boundless data.

Looming Privacy Concerns

As boundless personal data increases the utility of analytics to address business needs, it also runs the risk that the "creepy factor" will stop it dead in its tracks. Most of these data are collected and used without the consent of the individual. For example, personal data is collected from children as they traverse the Internet and then used for tailoring marketing messages to them. This concern about privacy is acknowledged but largely ignored, and the response is often to deny its existence. For example, one large technology firm said that "nobody's sitting around sifting through data on an individual basis and intentionally targeting specific individuals." And political campaigns have said that they are not "indiscriminately scooping up personal data on potential supporters." The facts would suggest otherwise. The data are very valuable and, while the gate is open, there are few restrictions on its use. But a few breaches of privacy might bring on a spate of consumer complaints and Congressional action.

Sociology, Not Technology

All the analytics sweet spots across the industries are successful because they complement and support important business needs and strengths. Excellent analytics methods can be developed in a bubble outside the realities of the business. But unless they are used to solve business problems, they collect dust and are an expense and not an investment. Making things happen/change/succeed is what business is about. It's not about the technology; it's the sociology. What is clear from the analytics sweet spots is that the bull's-eye for the value proposition for analytics is understanding business challenges and strengths and providing tools and expertise in the right way and at the right time to support the business.

Analytics Adaptations

The final stage in the discovery process is the translation of industry analytics sweet spots into potential adaptations in healthcare. It is hard to imagine, on the face of it, how banking is anything like healthcare. The task involves connecting dots between the industries to arrive at some revelation. It involves some logic, but is mostly about creativity. Creativity often leads to an epiphany.

The logical approach might proceed as follows: Both banking and healthcare were hit at about the same time with industry-shaking regulations for reform. And both possess a bountiful number of transactions. Indeed, banking is far ahead of healthcare in transactions processing efficiency, what with its pioneering innovations in self-service (ATMs) and its heavier regulatory compliance pressures. So one might infer that if banking does transactions better, the processes that support it should be considered for adoption by healthcare. It is certainly worth exploring, and we do so in the banking chapter, but it does not make the cut as a recommended adaptation. Replicating processes is not enough. We are looking for a sweet spot that is unique, addresses outcomes, and has a plausible bridge to addressing important concerns in healthcare as an adaptation.

Mostly, what translation requires is a creative leap. In the banking case, and reflecting on the industry strength in risk assessment, one ponders what it is about assessing customer capabilities and risks that might apply in healthcare. What is it? What emerged is that one of the most ingrained and intractable issues in healthcare is getting people to be active co-producers of their own health and thereby improve outcomes. People's behavior is the biggest influencer of health functioning followed by many social determinants. Can the risks, capabilities, and barriers to the fulfillment of doctor's orders, prescriptions, and plans that rest with the individual be measured and then managed better? This healthcare adaptation, called the Health Improvement Capability Score (HICS), is one of the seven adaptations that are distilled from the discovery process in the industry chapters of the book.

The seven healthcare analytics adaptations are listed in Figure 1.5 and are described in detail in Chapter 7, "The Top Healthcare Analytics Adaptations." The seven analytics adaptations cluster into three important areas of health and healthcare, including population health, patient engagement, and provider performance. The figure shows how each of the adaptations maps to the clusters. The adaptations for population health include obesity detection, well-being, and my dashboard. For patient engagement, the adaptations include radical personalization, and capabilities and support. Provider performance includes the adaptations of peer-to-peer and team centered.

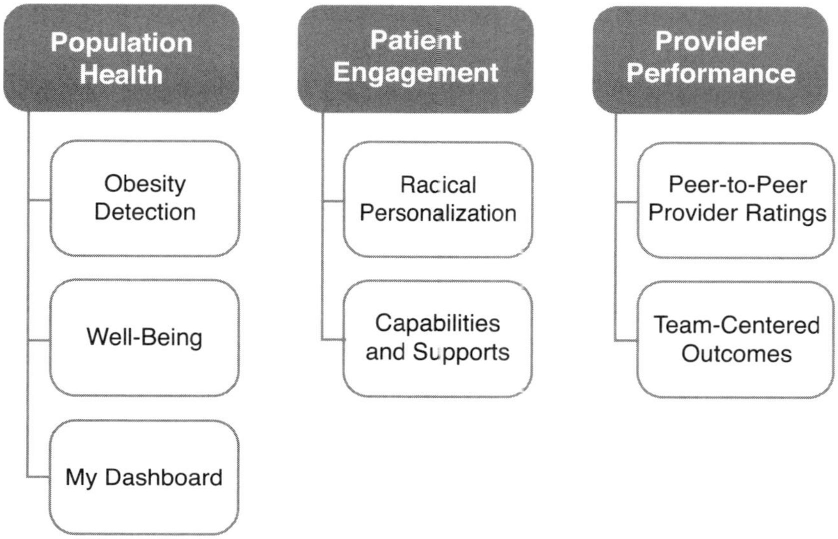

Figure 1.5 Seven healthcare analytics adaptations.

Putting Ideas into Action

We do not just leave the ideas hanging. Certainly, it is important to generate ideas and to show the way across the bridge to healthcare. But it is better to be a good steward and give the ideas their best shot to be put into action. A goal of the book is to provide a framework for the adoption of health analytics.

Ideas are fragile. Eggers and O'Leary think of ideas as seeds. Harkening back to a biblical parable, they note: "Some seeds land on rocky soil. Others are eaten by birds, and some sprout only to be choked by thorns. Only through a fortuitous combination of sun, soil, and water will a seed grow into a plant and bear fruit."[10] And progressing ideas to the endpoint of making a real difference is a long journey, and not all ideas deserve the passage.

The last chapter of this book provides a framework and tools for progressing an innovation from its early beginnings as an idea to the successful implementation of it into the operations of an organization. The Innovation Pathway (see Figure 1.6) depicts the six stages of this journey.

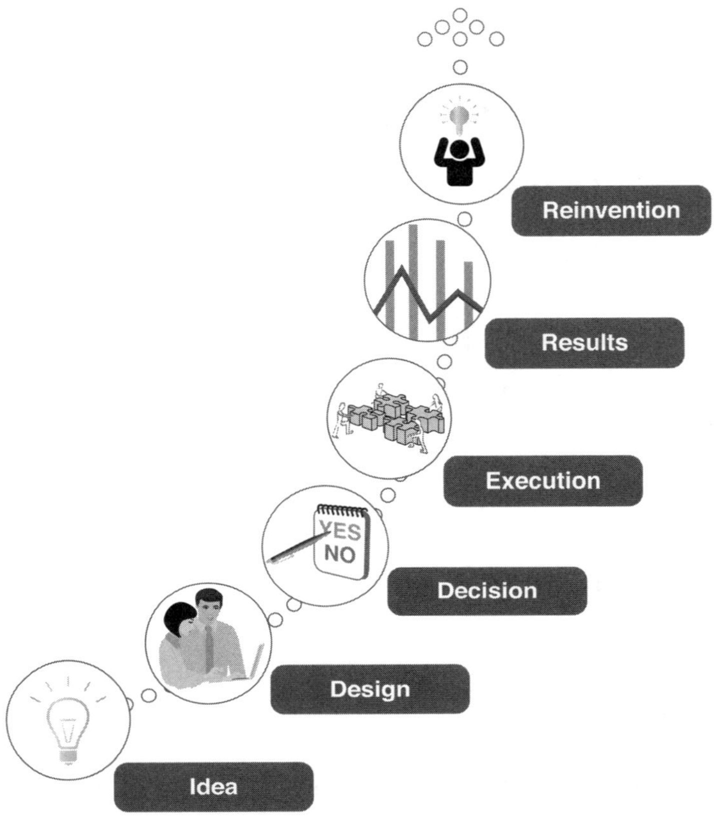

Figure 1.6 Innovation Pathway to Success.

We then concentrate on the adoption of innovations and propose an Innovation Adoption Factors Model (see Figure 1.7), which describes the six areas or domains that need to be considered and mastered in order to persuade individuals to make the decision to adopt an innovation. The Model is composed of two halves. On the left side is the idea stage and on the right is the design stage. The idea stage includes the domains of innovation receptivity, ideation maturation, and urgency/timing. The design stage includes the domains of the innovation's attributes, the organization's capabilities, and the readiness to position the innovation for adoption.

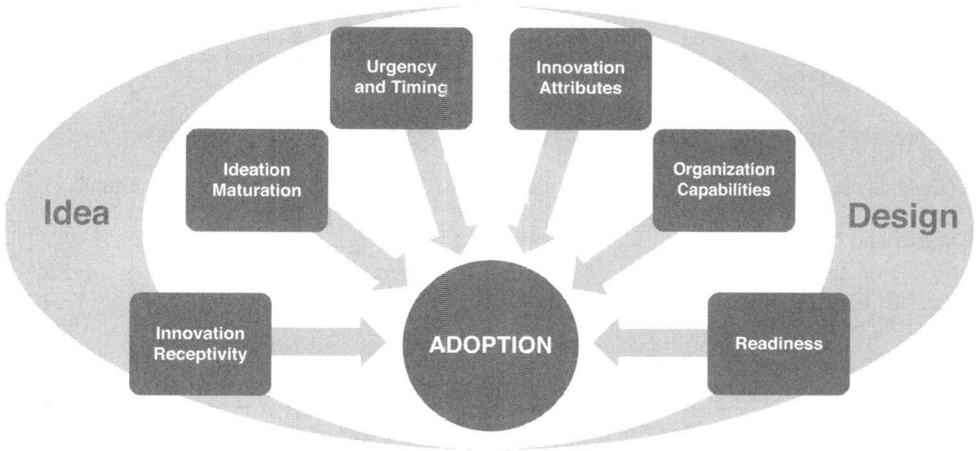

Adapted from Rogers on diffusion theory,[11] Eggers and O'Leary on getting things done in government,[12] Kingdon on agenda setting,[13] Hasenfeld and Brock on policy implementation,[14] Pressman and Wildavsky on program implementation,[15] Gladwell on the tipping point model of change [16] and Plsek[17] and Stacey[18] on complexity theory.

Figure 1.7 Innovation Adopters Factors Model.

We then break down the six domains into 18 factors and describe each of them. (See Figures 1.8 and 1.9 for a summary.) The factors become the basis for a guidebook that can be used for evaluation and planning for an innovation adoption.

The final chapter concludes with an application of the model to one of the healthcare adaptations proposed in this book, in the Health Improvement Capability Score, through a case study. The hypothetical health plan evaluated the 18 factors of the model and summarized the results in the Adoption Decision Dashboard (see Figure 1.10).

The evaluation yielded a "yellow" rating on the overall adoption decision index. A postscript reviews how the innovation was adopted after six months of hard work to bring the index to "green" and convince the executive committee to decide in its favor.

Innovation Receptivity	Ideation Maturation	Urgency and Timing
☐ Comfort with new ideas	☐ Quality of the idea	☐ Burning platform
☐ Survival pressures	☐ Confirmation bias	☐ Window of opportunity
☐ Norms and beliefs	☐ Cause and effect theory	

Figure 1.8 Idea generation domains and factors.

Innovation Attributes	Organization Capabilities	Readiness
☐ Relative advantage	☐ Individual's acceptance	☐ Prototype
☐ Compatibility	☐ Program theory/model	☐ Communications
☐ Complexity	☐ Skills and competencies	☐ Leadership
☐ Trialability		

Figure 1.9 Design domains and factors.

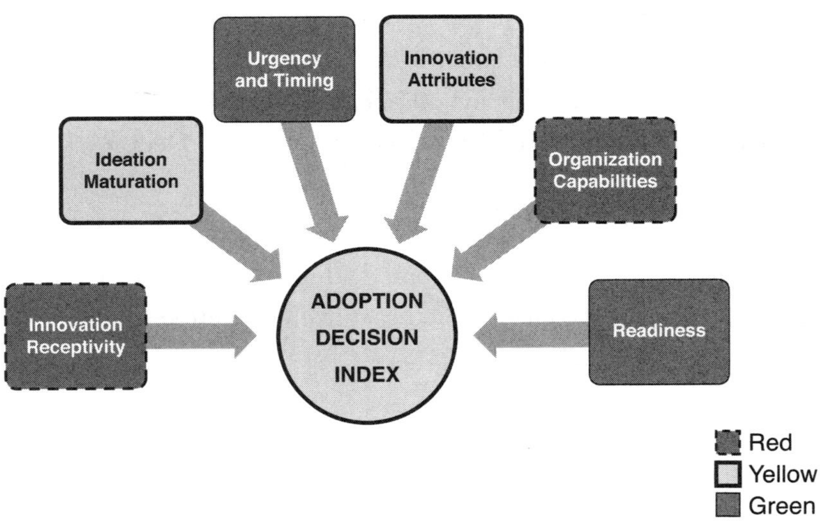

Figure 1.10 Adoption Decision Dashboard: Health Improvement Capabilities Score.

2

Healthcare

Although healthcare has sophisticated, scientific analytics in its research about the nature of diseases, its application to the delivery of care and running the business has lagged.

Introduction

The healthcare industry is wonderful, unique, complex, and intransigent. It is wonderful because it deals with the most important asset an individual has: one's health. We are distressed when we lose our health due to disease, accidents, or old age, and we are happy and thankful when we are relieved from pain, loss of function, and aloneness. However, the industry has many challenges in fulfilling these hopes related to achieving outcomes, efficiency, care delivery, and customer engagement. And it has many clear opportunities for breakthroughs in these areas and from other, less clear threats and challenges that can disrupt the business. The industry is very complex and has been intransigent to change on many fronts. Healthcare reform and market drivers are inducing change like never before.

Analytics can and must support the achievement of these business needs. Although healthcare has always had great scientific analytics in its research about the causes of diseases and the effectiveness of treatments, its application of analytics to the delivery of care and the running of the business has not been as successful. This chapter begins with a description of the unique and not-so-unique features and challenges of the healthcare industry. It ends with an overview of the current state of health analytics, including pointers to what can be learned from the business successes and related analytics sweet spots from other industries that are detailed in subsequent chapters.

It is a paradox that the success and failure of analytics in healthcare are related to NIH. The research conducted by the National Institutes of Health has led to remarkable discoveries. This book suggests that other industries can teach healthcare a lot through analytics and that a "not invented here" mind-set will blind the business of healthcare from opportunities within its grasp.

The Healthcare Industry Has a Unique Mission and Its Potential Contributions Are Great

A commonly expressed belief is "without your health, you have nothing." If people experience pain, symptoms, and loss of functioning for a prolonged period, they cannot work or fully enjoy the pleasures of life. And when these are relieved, through either medical intervention or time, people are very thankful. The psychologist Abraham Maslow, in his pyramid/hierarchy of human needs paradigm for self-actualization,[1] places basic physiologic needs at the base of the pyramid. Only after the basic physiological needs, such as food, shelter, water, and health, are satisfied can a person move up the pyramid to achieve safety, belonging, self-esteem, and the ultimate, self-actualization. Therefore, medical goods and services are viewed as necessities, like food, by most people and are valued highly.

What is health? According to the World Health Organization, health is a state of complete physical, mental, and social well-being and not merely the absence of disease or infirmity.[2] The healthcare system in the U.S. is primarily a sickness-focused industry. A multitude of distinct and unconnected systems address well-being in different areas, including public health, education, safety net, public safety, and many more.

Total U.S. health spending reached $2.7 trillion in 2011, amounting to about $8,500 per person.[3] It accounts for a substantial 18% of the economy (GDP). The rate of growth in spending abated during the Great Recession years of 2009–2011 at 4% per year, probably because of lower consumption and, perhaps, improvements in the delivery of care. However, when put into perspective over the long term, health expenditures have risen by a factor of ten over the past 20 years, and it is doubtful that this recession reprieve will continue unless there is a substantial change in the underlying fundamentals. The U.S. continues to spend much more of its GDP on healthcare compared to all other countries—indeed, almost twice as much as other rich countries.

Although the U.S. government refers to national *health* expenditures in its statistics, the vast majority of expenditures pertain to health *care*. For example, annual hospital expenditures are approaching $1 trillion and physician services are more than $0.5 trillion.[4] These two categories represent more than half of all expenditures. The next-largest area is prescription drugs, at 10% of expenditures. The remaining nine categories, including nursing home care, home health, government-sponsored health insurance, and investments in research, compose the remaining 39%. There is no budget category for prevention. Very little is spent on public health—only $82 billion or 3% of total expenditures—and this has not changed much over the past 20 years.[5]

Because of the necessity of providing care when people are sick, the seemingly intractable difficulty in controlling medical costs, and the collection of low tax revenues for governments resulting in flat budgets, healthcare has sucked up a disproportionate share of government budgets. For example, in the Commonwealth of Massachusetts, state spending for health coverage increased 59% over the ten-year period ending in FY 2011. All other government human service budgets decreased. The hardest hit was public health, with a 38% decline. Mental health was down 33%, education down 15%, housing down 23%, human services down 13%, and public safety down 11%.[6]

As we will discuss shortly, the U.S. is not getting a good return on its heavy investment in healthcare as compared to other wealthy countries. Indeed, the outcomes across many domains are the worst in the U.S. This might be because the U.S. is really *not* investing as much in *health* as its peer countries. In fact, it ranks tenth.[7] Elizabeth Bradley and Lauren Taylor looked at spending on social services such as unemployment benefits, pensions, family support services, and rent subsidies and added it to healthcare spending. When they compared the U.S. to 30 other industrialized countries on this combined measure of health spending, they found that the U.S. does not spend the most by a long shot.[8] It spends about 29% of GDP on this bundle of health services. Other countries, such as Sweden, France, the Netherlands, Belgium, and Denmark, spent 33% to 38% of their GDP on health.

Additionally, they noted that for every dollar spent on healthcare in the U.S., it spends an additional 90 cents on social services. Yet in the comparison countries, every dollar spent on healthcare is complemented with $2 on social services. So a greater weight is placed on social services in these healthier countries than in the U.S.

As we will discuss shortly, the research is abundantly clear that social determinants are far more important in the production of health than the delivery of healthcare. The vast majority of physicians agree that unmet social needs are a direct cause of poor health. However, proposals for U.S. government spending for social programs bring out strong political opposition from the Right concerning support for a "socialist state" at the expense of "individual freedoms." And individual freedoms do not often translate into behavioral changes to improve health. For example, personal behaviors directly related to obesity and violence are unique features of the American way of health and death.

First and foremost, health is at the core of the essence of life and is a vital foundation for the attainment of happiness and well-being in society. (The other industry that approaches these aspirational goals of well-being and achievement of human potential is the education system.) The mission of the other industries covered in this book, including retail, banking, politics, and sports, is relatively inconsequential:

- Retail makes, distributes, and sells goods, many of which are vital to well-being, including food, clothing, and transportation. The retail industry depends on and fosters a consumption society, and much of what is sold is nonessential, even frivolous. Americans love their shopping, derive a good deal of pleasure from it, and drive the economy accordingly through personal spending. Clearly if a wide spectrum of goods were not available, society would grind to a halt, but this is very unlikely in a vibrant market-oriented economy. The real limiting factor to achieving well-being from retail is the availability of money, which mostly comes from jobs, social supports, and good health.
- Banking facilitates the handling of money so that people can address their economic security. Retail banking facilitates the individual's need to move,

store, use, and invest money. Banking's primary role is to handle transactions. But, as described in the preceding point, economic security is driven more by employment, social supports, and health status.

- Political campaigns boil down to voting. Voting is important for the full expression of democracy in a society in its choice of its representatives who subsequently make government policy. Voting is a fundamental civic right. But if an individual did not vote—and just under 50% do not vote in a presidential election—it would not limit one's capability to live a satisfying life. If particular segments of society were not allowed to vote, as was the case with women and African Americans until the twentieth century, this could lead to social unrest and a poorly performing democracy and potentially to devastating economic problems. But at the individual level, voting does not contribute significantly to day-to-day happiness and well-being.
- Sports is about entertainment. Although many people spend lots of their leisure time watching sports and getting quite innervated about March Madness and the Super Bowl, it is primarily about leisure. Without sports, there would be alternatives to occupy people's need for relaxation.

So health is a vital resource for people to survive and achieve well-being. Its purpose and mission are superordinate to other industries in terms of its contribution to society. Yet despite the fact that these other industries do not deal with life-and-death matters like healthcare, they can teach healthcare a lot. These other industries are very successful in business and have honed certain skills, techniques, strategies, and analytics that are often more refined than those in the healthcare industry.

The Healthcare Industry Has Major Challenges

The U.S. healthcare industry has five major challenges: suboptimal outcomes, voltage drop from science to the delivery of care, the need to move beyond a sickness system to achieve population health, inefficiency, and low customer engagement. These are described in the following sections.

Suboptimal Outcomes

It has been well known for quite some time that the U.S. does not achieve a good return on its extremely high expenditures for healthcare when compared to other industrialized countries. The Organization for Economic Cooperation and Development (OECD) published a report, *Health at a Glance in 2011*,[9] that ranked the U.S. at 28th among OECD countries on the most important measure of healthcare outcomes, life expectancy at birth. The U.S. was ranked lower than Slovenia and Chile and just above the Czech Republic, Poland, and Mexico. The U.S. gained 8.3 years in life expectancy over the past 50 years, which is a substantial gain when compared with life years gained in previous decades. However, when compared to other countries like Japan, the 8.3 years is just a bit more than half of what Japan gained. Life expectancy in Japan today is 5 years more than that in the U.S.

A report published in 2013 by the National Research Council and Institute of Medicine, *U.S Health in International Perspective: Shorter Lives, Poorer Health*,[10] supported and extended these findings. The panel documented that the U.S fares the worst when compared to 17 high-income countries on nine critical health domains:

1. Adverse birth outcomes: infant mortality rate
2. Injuries and homicides: deaths from motor vehicle crashes, non-transportation-related injuries, and violence
3. Adolescent pregnancy and sexually transmitted infections
4. AIDS: highest incidence (HIV second highest)
5. Drugs- and alcohol-related mortality: years of life lost
6. Obesity and diabetes: highest rates and prevalence
7. Heart disease: death rate second highest
8. Chronic lung disease: higher than European countries
9. Disability: higher prevalence of arthritis and activity limitations

The panel also noted a surprising and disturbing finding that life expectancy is getting worse for people in the age group under 50 years in the U.S. as compared to other countries. Over the two-year period from 2006 to 2008, the number of years of life lost for this large group was higher than that in the other 16 countries.

For men it was almost 1.4 years and for women it was 0.8 years—*in just two years*. Recall that it took 50 years for the U.S. to gain 8 years. Further, when the panel decomposed the data to understand what conditions were contributing the most to the life years lost in the U.S., more than half of the loss was attributed to causes not usually associated with healthcare, including violence and accidents. Note that the age-adjusted mortality rate for violence in the U.S. is much higher than that in all 17 countries and is six times higher than the lowest-mortality countries, including Japan, Austria, Norway, Switzerland, Germany, Denmark, Sweden, and Spain. The next-highest category contributing to the difference in loss of life years was perinatal conditions (mostly infant mortality), at 13%.

They suggested that in addition to known contributors to poor health performance, including lack of insurance, social and economic disadvantages, and race and ethnic disparities, there is a "health-wealth paradox" of a pervasive disadvantage that affects all Americans, including even the relatively well-off who do not smoke and are not overweight.

Voltage Drop from Science to the Delivery of Care

Important outcomes directly related to the delivery of healthcare are also the worst in the U.S. when compared to peer countries. One such measure is mortality amenable to healthcare. Included in this mortality rate are diseases with a well-known clinical understanding of its prevention and treatment, including ischemic heart disease, diabetes, stroke, and bacterial infections. In other words, the science is very clear on what needs to be done, and the premature mortality rate (for people under 75 years of age) is an important indicator of the success of executing on the science on what's known. Figure 2.1, adapted from the Commonwealth Fund's *National Scorecard on U.S. Health System Performance 2011*,[11] compares the mortality amenable to healthcare rate compared to the averages of 15 other wealthy countries, which are nearly identical to the WHO research given previously. The U.S. rate of 96 deaths per 100,000 lives is almost 40% higher than the average of the best five countries, including France, Australia, Italy, Japan, and Sweden, at 59 deaths per 100,000. The U.S. is almost twice as high as the country with the lowest rate, France, at 55 deaths. Similarly, the U.S. rate is higher than the average of the middle tier of countries (69) and the countries with the highest rates (79). Further, the rate of improvement in the U.S. death rate over 10 years, between 1997 and 2007, is 20%, which is considerably less than the average for all other 15 countries, at 32%.

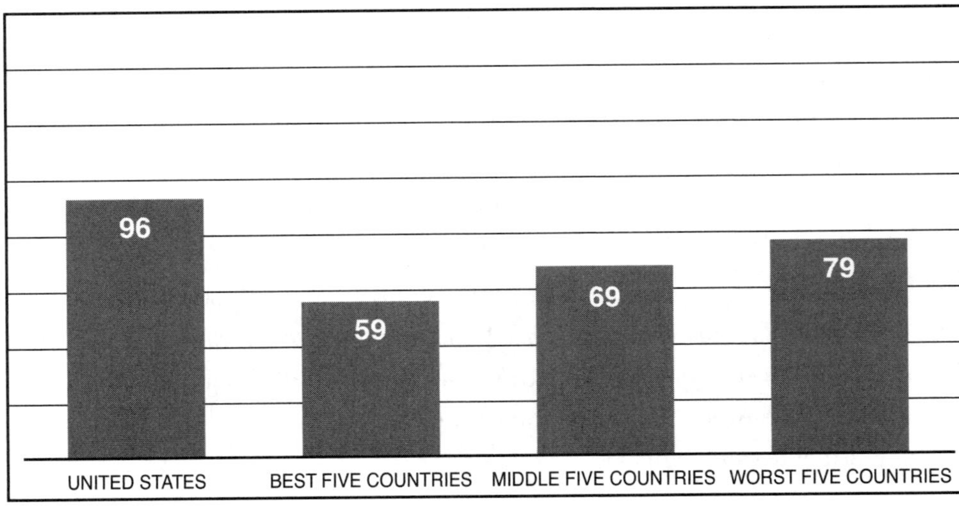

Source: Adapted from The Commonwealth Fund, 2011.

Figure 2.1 Mortality amenable to healthcare: premature death rates per 100,000.

The Commonwealth Fund also computes performance scores for five overall dimensions of a high-performing health system, including healthy lives, quality, access, efficiency, and equity, as well as an overall score. The U.S. gets a score of 70 out of 100 for "healthy lives" (the dimension most indicative of outcomes) and this score has fallen by five points since the 2006 Scorecard. The lowest score among these five domains is for efficiency, at 53. The overall score across all five domains is 64, which has been falling somewhat over the past 5 years.[12]

The voltage drop from the scientific bench to the clinical bedside (from research to practice) was documented in a now-classic study authored by Elizabeth McGlynn and her colleagues at the RAND Corporation, "The Quality of Health Care Delivered to Adults in the United States."[13] The research addressed the clinical adherence to recommended processes of care for 30 acute and chronic illnesses, as well as preventive care, with 439 indicators. Overall, the results showed that patients received recommended care about 55% of the time. They concluded that these deficits in the provision of recommended care "pose serious threats to the health of the American people."[14]

The Need to Move Beyond a Sickness System to Achieve Population Health

Health outcomes are a function of many factors. And the least important, despite its high cost in the U.S., is healthcare. The determinants of health have been well researched and documented in the "social determinants" literature. As Steve Schroeder said in his excellent Shattuck Lecture on improving the health of the American people, "even if the entire U.S. population had access to excellent medical care—which it does not—only a small fraction of [premature deaths] could be prevented."[15] The best distillation of the available data by McGinnis et al.[16] indicates that the single greatest challenge and opportunity to prevent early death lies in modifying personal behavior. Figure 2.2 indicates that personal behavior accounts for more than 40% of health functioning, and shortfalls in healthcare account for only 10%. That is a fourfold difference. And the combination of behavior patterns with social and environmental factors adds up to almost two-thirds of all premature deaths.

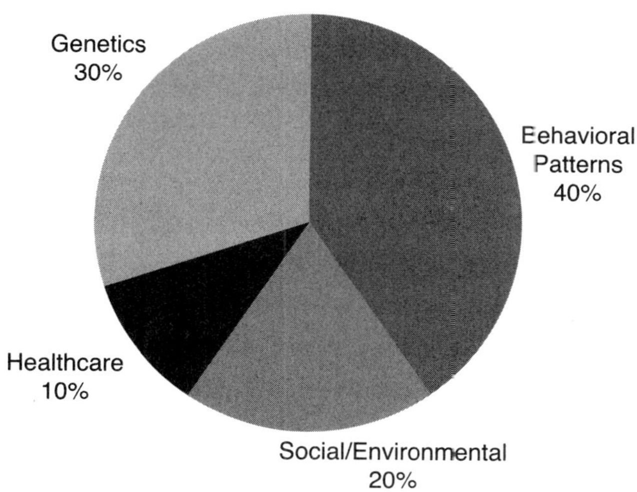

Source: Adapted from McGinnis et al., 2002.

Figure 2.2 Determinants of premature death.

Deaths attributable to behavioral causes amount to more than a million a year, including smoking at 435,000; obesity and inactivity at 365,000; alcohol at 85,000; motor vehicle accidents at over 43,000; and guns at 29,000.[17] Changing behavior is the unrealized holy grail of producing better health, and it might be more influential in improving the health of the population than technological innovations.[18]

Social circumstances are also more influential than healthcare in determining health. And these are hard to change (including improvements in education, transportation, housing, and food). Schroeder suggests that a basic U.S. preference to value entrepreneurialism over egalitarianism, which results in inequality in income, education, and more, has definitive unintended health consequences.

What is clear is that the factors associated with good health go way beyond the traditional functions of the healthcare system. There have been accounts of doctors prescribing stimulants like Adderall and Concerta to underprivileged students to help them learn better and compete with students with more stable social environments, although these students do not have an A.D.H.D. diagnosis.[19] This is an understandable, but unsavory, way to correct social inequalities. There are programs that work to supplement the fixation on sickness by providing a means for doctors to give "prescriptions" for transportation and healthy food. And the movement toward population health through accountable care organizations (ACOs) will place a greater emphasis on prevention and effective treatment.

The responsibility for the nation's health well-being is dispersed through many public and private health systems and other social systems. It's hard enough to get doctors within a practice or hospital or system to communicate with one another. Coordination across all these social and health systems seems almost inconceivable. The good news is that the U.S. does have experience and successes with organizing communities for better health as it did during the 1960s and 1970s and especially through the Great Society programs. Perhaps there will be a cyclical swing in U.S. politics back to a more progressive orientation to improving well-being for all through community-wide initiatives.

However, since social factors account for 15% of health and are extraordinarily complex and hard to change, a more proximate and potentially better return on investment would be in the primary area of behavior change and how to fully mobilize each individual's ability to co-create their own health. We will return to this topic a number of times, through a variety of industry lenses, and with different analytic tools, as a possible breakthrough in improving health.

Inefficiency

Of all the performance metrics of the U.S. health system, efficiency is the worst. The Commonwealth Fund National Scorecard in 2011 gave it a score of 53 out of 100. It looks at indicators covering potential overuse; preventable emergency room use, hospitalizations, and readmissions; use of electronic medical records; and costs of care and mortality for specific conditions such as hip fractures and colon cancer and for multiple chronic diseases. Here are a few examples of worst performance metrics of the U.S. healthcare system relative to those of other countries:

- More duplicate medical tests
- More tests or medical records not available at time of appointment
- ER visits for conditions that could have been treated at home
- Highest insurance administrative costs
- Lowest use of electronic medical records (except for Canada, 2009)

The data on variation on costs and outcomes has also been well documented by the Dartmouth Atlas[20] on a wide variety of measures, including hospital and physician utilization, surgical procedures, intensity of care, care for chronic conditions, and end-of-life care. The Dartmouth data looks at regional and hospital variation and has been a major data resource to spotlight variation in clinical practices as it relates to cost and quality. Atul Gawande's use of the Dartmouth data to expose waste and abuse of the fee-for-service system in McAllen, Texas, in his classic *New Yorker* article "The Cost Conundrum"[21] ignited concerns about overuse in the U.S. healthcare system.

Finally, when the efficiency of the healthcare system is compared to the efficiency of other systems that address societal needs, such as government, finance, communications, transportation, and more, healthcare is rated the worst on two metrics. The first metric is a measure of system inefficiency as a percentage of total economic value of the system. Healthcare gets a score of 40%, higher than any other system. For example, on this metric communications is at 22%. The other measure is improvement potential as percentage of system inefficiency. Again, healthcare ranks lowest, meaning it has the greatest improvement potential with a score of 35% as compared to government at 27%.[22] This corresponds well with many reviews that peg waste in the U.S. healthcare system at 30% or more.[23]

Low Consumer Engagement

If behavior is vitally important to improving health, engaging people is the first step, and then providing a good customer experience motivates, sustains, and satisfies them. But people have chosen, by and large, not to be engaged in healthcare despite the fact that they need to become genuine co-producers of health in partnership with their provider, payer, pharmacy, and other health system counterparts.

Customer experience can be measured. Forrester Research rates customer experiences across industries.[24] It defines customer experience in terms of (1) meeting needs, (2) being easy to work with, and (3) enjoyability. Its 2011 survey results show that health plans/systems are dead last in the rankings among 14 industries, even lower than credit card providers and Internet service providers. Kaiser Permanente is the highest-rated health plan/system surveyed, with a ranking of 75 out of the 133 companies across all industries rated. Medicare is 102 and United Healthcare is close to the bottom of the list at 131.

Not surprisingly, retailers and hotels rank highest. They understand the need to satisfy customers in an industry that depends on repeat business. Also, insurance providers in other industries do much better than the healthcare insurance industry and rank fifth with an average score of 72%, as compared to health insurance at 51%. Their direct connection with the customer, as contrasted with the indirect connection through intermediaries in health insurance, might say a lot about the relatively higher regard paid to customers and the appreciation voiced by customers in their ratings.

Customer engagement going forward might get complicated in all industries. Because of advances in mobile phones, the influence of social media, and the consumer demand for simplicity, convenience, and transparency, there might be significant social change brewing that will turn the tables on conventional ideas about customer experience. Other industries are responding to this...healthcare is lagging.

Summary: Monetizing the Challenges

In summary, the challenges to the U.S. health system in relation to outcomes, efficiency, and customer engagement are huge, and the opportunities for improvement are also huge and could be the basis of competitive differentiation. The costs in terms of lives lost and quality of life are most important. However, if one were to monetize the value of closing the gap between present performance and benchmarks, the rough estimates are as given here:

- **Outcomes**

  If the U.S. had the same mortality rate amenable to healthcare as the average of France, Australia, Italy, Japan, and Sweden, it would be about 40% less and save 118,000 lives a year. At the U.S. government's estimate of the value of a life at about $7 million,[25] this loss of life amounts to $826,000,000,000, or close to one trillion dollars.

- **Efficiency**

  If healthcare eliminated all waste (midpoint estimate of 34%), it would save almost one trillion dollars a year, according to Berwick and Hackbarth.[26] If healthcare were as efficient as the most efficient system, communications, it would be at least 50% more efficient than it is now, and this would add a half trillion in savings.

- **Customer Engagement**

  If the (average) health insurance plan performed like the (average) retailer, there would be a 61% improvement. Better customer engagement would increase market share and related profits. If one assumed a very modest increase in market share of only 10% related to significant improvements in customer engagement, this would result in the capture of 2.5 to 5 million new customers as a result of the increased number of people projected to be insured through the Affordable Care Act (25 to 50 million) and monetized at a total premium value of $28 billion dollars (based on average individual premiums of $5,650 in 2012[27]). If 10% of the existing pool of people with insurance (100 million), or 10 million people, shifted plans, this would result in more than $50 billion.

The purpose of this monetizing exercise is to demonstrate the order of magnitude of possibilities, not to provide exact projections. Indeed, the savings are based on reaching the full potential relative to benchmarks, which is seldom accomplished given the complexities of implementation. However, if the healthcare system reduced waste by just half, that would amount to $500 billion. Similarly, understanding the value of life lost as a result of not executing delivery well is important even if it resulted only in achieving half the goal or $500 billion.

Healthcare Is a Very Different Industry in Many Respects

Healthcare is complex, much more so than most other industries. It is based on sophisticated research, is practiced by highly trained professionals, uses high-tech equipment, and delivers its products and services through multiple settings. It is often uncoordinated and unconnected. It is headed by an elite profession. It does not follow many of the market fundamentals of other industries.

The primary focus of most industries is on the customer for the simple reason that without the continued support of customers in buying the industry's products and services, businesses could not compete and would go out of business. However, healthcare does not think of people as customers. It uses various terms to describe them, including patient, member, consumer, and citizen. This is partly understandable because the people who get treated, enroll in health plans, or buy medications are not regarded as the real customers.

A customer is defined as a person or an organization that buys goods or services from a store or another business. In healthcare, employers are the buyers of health insurance. Doctors contract with insurers. Pharmaceutical benefit managers make deals with employers. People do buy a lot of health and wellness products, and many go to doctors a lot and a minority are treated in hospitals. But when it comes to purchasing medical services, they usually pay only a fraction of the cost, use other people's money (insurance), follow employer/government benefit rules on plan choice (if any) and "allowable" services, and mostly "buy" what the doctor orders.

In addition to limited options—for example, one can only go to the prescribed in-network hospitals or doctors—people cannot really make a choice at the point of service because they are often sick or in an emergency. For example, take the case of the person who is brought to the ER unconscious and the hospital assigns an expert trauma surgeon to operate. But the surgeon is out of network, and the patient, without having any choice, is billed a lot more than if the doctor were in network.

In addition to limited choice and fake money, both of which seriously compromise the basic requirements of well-functioning markets, consumers have little information on cost and quality to help with their purchasing decisions. Although public reporting of performance metrics in healthcare are forging the way for greater availability of quality and cost information to drive accountability and better performance, transparency in the healthcare industry is subpar when compared to transparency in other industries. Most people do not use the information for making choices, and most experts would agree that transparency's true purpose is to publicly expose information to providers and insurers so that they can compete more vigorously. This is very different when compared to other industries. In retail, for example, customers do "show-rooming." A customer goes to a store, checks out the product, pulls out the smartphone and compares prices from all other stores, and then makes a decision based on readily available price and quality information. It seems inconceivable that show-rooming will ever become a reality in healthcare. But it must.

The costs and outcomes of healthcare are largely driven by the decisions of intermediaries, not individuals, even though the latter pay for it directly out of pocket and indirectly through taxes and lower wages. As previously noted, employers and health insurers make many decisions on behalf of people. But the most important intermediary is the doctor who is responsible for most decisions related to a patient's healthcare and directs about 80% of the costs related to care.

Even pricing is different in healthcare. Usually price is determined based on input costs, perceived value, and what the market will bear relative to the competition. In (commercial) healthcare, the price of services is negotiated between insurers and providers. Prices for the same CPT code for very common services can vary significantly across insurers and providers. For example, in a study by the State of

Massachusetts[28] on statewide payments for services across hospitals, the prices paid for an appendectomy (DRG 225) varied by more than 11-fold for a severity-1 stay and 16-fold for a severity-2 stay. Prices paid for each of the other selected DRGs varied less significantly, but in no instance did they vary by less than 300% statewide. Providers might proclaim that they charge more because they have better quality. But that is seldom the case. One of the major drivers of higher prices is the market and brand power of a provider group, sometimes referred to as "brand blackmail" that can extract higher payments from insurers.

Healthcare is heavily regulated and there are many business compliance requirements. Additionally, some features of the ACA will induce transformation of the industry as no market forces could possibly do. For example, the Affordable Care Act changes the marketplace for health insurance by legislating health insurance exchanges, making certain risk management practices illegal; for example, insurers cannot deny insurance because of preexisting conditions, and states will review health insurance rate increases for reasonableness. It also opens up the marketplace such that individuals can buy insurance from insurers directly from the exchanges. The CBO estimates that 25 million Americans will buy insurance on the exchanges. Others estimate that more than 70 million will eventually do so if an estimated 30% of employers stop offering coverage.[29] This will require insurers to focus on "retail" customers, their members, and prospective members, in addition to selling to (wholesale) employers. Additionally, there are strict rules on privacy and sharing data, as with HIPPA. And there are regulations on bringing products to market that require evidence of safety, efficacy, and fair advertising.

Finally, all industries have their own norms and cultures. What might pass as "pulling the wool over the eyes" in retail marketing might not fit well with the perceived culture of the caring professionals in healthcare. However, it could be argued that the relentless provision of more and more diagnostic tests leading to unnecessary procedures is driven more to improve "revenue enhancement" than to improve the patient's health. Similarly, the relentless onslaught of advertising for prescription drugs for newly profitable conditions—now the favorite is low testosterone in men, last year it was toe fungus, and of course there is the ever-popular erectile dysfunction—uses perfectly designed images and messages to sell lots of pills.

There are other aspects of the healthcare industry's culture that are unique. Most people who go into the healthcare professions are more caring and dedicated to improving society than are those who enter careers that are more directly focused on commerce and profits. Additionally, there is a strict hierarchy of power and influence, from the doctor downward There is a stubborn orientation to be provider-centric rather than patient/customer-centric on straightforward issues such as convenience and shared decision making. There is an inherent autonomy in the profession of doctoring that makes it particularly impervious to change. And there are special peculiarities of the business of healthcare, which are emphasized later.

Healthcare Is a Very Similar Business on Some of the Fundamentals

Healthcare is big business with national revenues for doctors and hospitals at almost $1.5 trillion in 2013. The size of the largest healthcare companies is impressive. Kaiser Permanente, an integrated delivery system and health insurance plan, has almost 9 million members/customers, 173,000 employees, 16,658 physicians, 37 hospitals, 611 medical offices, and operating revenues close to $50 billion in 2011.[30] The largest health insurance company is United Healthcare Group, with more than $100 billion in revenues. The largest life sciences company is Pfizer, at $50 billion in total revenues.

However, despite these behemoths of the industry, most care delivered by the nation's nearly one million doctors is performed in solo practices, and 37% is in small group practices of two to five physicians.[31] Hospital care is delivered in nearly 6,000 hospitals. And the vast majority of care delivered in hospitals (87%) is in community hospitals with an average size of about 150 beds. Additionally, fewer than a third of the nation's 6,000 hospitals are in a network that works together with physicians, insurers, and other community agencies to provide coordinated care.[32]

The healthcare industry has been characterized as a cottage industry by some. A cottage industry refers to establishments that create tailored products and services, usually in the home or in shops, and by families or solo craftsmen. The cottage approach is very different from a mass production approach. As mentioned earlier, much of the healthcare industry is composed of doctors in solo or small practices, doing their work in small offices, and in relatively small community hospitals. The industry is not coordinated or integrated to any large degree and is led by "dedicated artisans" (doctors) who overvalue autonomy and eschew any form of standardization that would certainly be required for "productionalized" medicine.[33] This autonomy, reliance on intuition, and reluctance to seek feedback for improvement blunts the use of disciplined science in determining whether to embrace clinical guidelines or to adopt efficient business processes. For example, were it not for the meaningful use payment of up to $63,750 from Medicare and Medicaid to implement electronic health records (EHRs), most doctors would demur because they see it as a distraction to their office efficiency. This cultural DNA of practitioners defines the character of the industry and influences the performance of the business.

On the other hand, many (doctors) would argue that the business of healthcare has turned to market-oriented solutions implemented through corporate capitalism over the past 25 years or so. This, of course, is quite different from a cottage industry! This turn toward market-oriented solutions and away from government-oriented solutions happened in the 1980s as Americans, or at least the majority of their elected politicians at the time, decided that healthcare should be run by the same economic and business strategies that drive other industries. It was widely held that costs and medical inflation were out of control, that government solutions had failed, and that it was time for business solutions that were successful in other industries to be tried in healthcare.

For doctors, this has resulted in many forms of oversight, including contracts with managed care companies that resulted in monitoring of their economic and clinical performance along with increasing pressure to comply with standard clinical and business practices. Some market-oriented approaches became unpalatable to Americans, including policies on preexisting condition exclusions and high annual insurance premium increases. Thirty years later, government stepped in to reinstate government regulations to shape the insurance market with the Affordable Care Act. However, the healthcare industry is still largely a market-driven industry, despite the fact that Medicare and Medicaid payments by both the federal and state governments are about the same as employer and household payments for healthcare.[34] Despite the funding source care is mostly provided in the private sector and through private health insurers

Previously, we concluded that the healthcare system continues to be very inefficient and ineffective relative to other systems/industries and relative to its own standards. On the other hand, the healthcare industry overall is profitable and has done well for its investors. Healthcare (including insurance and care delivery) ranked 14th in profitability among 35 top industries with return on assets of 3.7%, and 19th in terms of profit growth at 8.4%, in 2009.[35] (The most profitable industries were mining and oil production at 19 8% and pharmaceuticals at 19.1%.) The largest healthcare company, UnitedHealth Group, realized more than $5 billion in profits in 2011, an 11% increase from 2010.

What are we to make of this paradox? How can an industry be so successful from a business point of view yet so subpar on addressing key business challenges? It depends on what the business objectives are. It is not the responsibility of business, per se, to attend to society's needs. It survives by responding to business needs, which are not always aligned with society's needs. It stays in business to the extent that it satisfies shareholder/investor expectations to achieve good earnings. First and foremost, business wants to do well for its investors, and secondarily it wants to do good for society.

There are many examples of healthcare business practices thwarting the public good, including the aggressive lobbying of Congress to extend patent or pricing protections for a pharmaceutical firm at tremendous expense to society,[36] the perpetuation of a fee-for-service payment system that often produces more volume and revenues than effective outcomes, and the marketing of "me, too" drugs that provide little incremental value while ignoring research and development on orphan drugs to benefit the suffering of people with rare diseases. Governments attempt to step in to address the public good and create a fair economic playing field for the market to achieve societal goals. For example, the State of Massachusetts has enacted global payments for all state-funded programs and expects the private sector to follow suit. The fact that market-driven industries, all of them, do what's important to their shareholders first is not an earth-shattering insight. But what is paradoxical is why some healthcare companies have not taken on the challenge to bridge the gap to provide breakthrough performance to improve care and thereby achieve competitive differentiation.

All businesses in all industries survive and can thrive by creating a competitive advantage in the marketplace. The core of the advantage is the ability to meet customer needs more effectively with products and services that customers value more highly or to meet needs more efficiently through lower costs and prices to customers. Companies like Walmart and Southwest Airlines are good examples of low-cost providers. Good examples of companies that have differentiated themselves on the basis of high quality, added performance, design, or technical superiority include Apple for innovative products and design, Rolex or Armani for brand prestige, and Mercedes for engineering prowess.

In healthcare, price is very important when it comes to insurance because employers tend to pick the lowest-cost options (or subsidize employees' co-share of premiums for the most efficient plans so that they follow suit in selecting low-cost plans). Employees also tend to use price as the major factor in choosing a plan if they have any choice at all. They presume, probably rightly, that there is no true "quality" differentiation among healthcare insurers in a given market area. When it comes to providers, "prices" are negotiated with insurers largely based on brand "power." Some healthcare providers get better prices simply because they can... because they have market power; for example, a provider might do so by threatening to not accept an insurance carrier and thereby deny access for its members to its highly branded hospitals unless its price demands are met. Utilization of providers by patients is not usually a function of price or quality since neither is known by the patient, although there is wide variation in what insurers pay providers and what providers bill patients. Choice of hospital and doctor is usually constrained by network configuration. Often, patients simply pick the doctor who lives closest to them and speaks their language.

There are many shifts underway in healthcare, driven by both government and market forces, which can be exploited to achieve market differentiation that can supersede competition on price alone:

- **Retail transformation:** Healthcare reform and other forces are requiring insurers to change their orientation on who the customer is. They will need to go from a wholesale business-to-business approach with employers to a retail business-to-customer approach with individuals. This will place emphasis on pleasing the retail customer with good service, transparency on price and performance, and radical personalization to satisfy the customer. These have not been strong skills of health insurers. Winners can emerge to earn greater market share.

- **Delivery system changes:** The movement toward ACOs is moving forward with the Centers for Medicare & Medicaid Services (CMS) offering a number of pilot programs. Related to ACOs are medical homes, payer and provider blending/merging, and full actualization of patient centered care. These innovations can lead to efficiencies to reduce price and improve quality.
- **Payment changes:** The move to global payments is very important to diminish the wasteful sequelae of the fee-for-service system. If the theory is right, and it is implemented well, it should lead to efficiencies to improve outcomes and lower costs.
- **Focus on outcomes:** Related to payment and delivery-system changes is a focus on outcomes. Can you imagine if one integrated delivery system could state with certainty that it can prove that they provide better outcomes than their competitors for various conditions? This is different from the publicly reported scorecards that include arcane measures, at least in the eyes of regular people, such as the rate of administering beta blockers for heart attack.

In addition to these known clinical and business opportunities in healthcare, there are future unknowns that can fly under the radar screen and can cause considerable disruption. These were referred to in Chapter 1, “Overview,” as Kodak moments of 2012 and indicate the need to scan for game-changing technologies that can either make the business obsolete or make it absolutely untouchable from a competitive differentiation point of view. These potential game changes are as follows:

- **Behavior change:** The greatest unmet challenge in health production is individual-driven behavior change. Closing this gap will result in great strides in prevention, care management, and outcomes.
- **Democratization:** Healthcare is provider-centric. Web 2.0 and social media change the locus of control, emphasizing the power of networks, and

how care can be purchased, evaluated, and modified. This might provoke the need to be people-centric and for organizations to take their lead rather than prescribing procedures and rules to follow.

- **De-medicalized health systems:** Medical care is a small contributor to health. New organizational forms of health production will be induced as cost pressures grow and outcomes lag. In addition to medical homes, there might be health houses.
- **Predictions:** Use of huge volumes and different, but highly relevant, types of data from within and outside the industry will lead to significantly better predictions to inform an array of strategies and decisions.
- **Mobile:** People have grown a new bodily organ and it's called an iPhone or a smartphone. It can have amazing functionality. It epitomizes consumers'/people's need for convenience, simplicity, immediacy, autonomy, and technology that works for them.

The Current and Future State of Health Analytics

To exploit opportunities for market differentiation and to avoid Kodak moments, analytics needs to meet the challenge. And although there has been great progress in analytics in healthcare, it is far behind other industries in supporting business objectives. What's holding it back the most is the digitization of the industry. Other industries, such as retail and banking, have perfected the digitization of their business transactions and reap the benefits from analytics. It is definitely true that healthcare has much more complex and varied types of transactions, for example, billing, clinical, and plan enrollment. Additionally, the industry is very fragmented. If the different segments of the system do not communicate well, it makes for a nearly impossible task to build an analytic infrastructure with them or around them to foster communication and integration.

Just as in Maslow's hierarchy for self-actualization, a business must mature through a series of stages before it can reach its full potential through analytics. At its base, the "physiologic" needs (like food and water) need to be secured first. In healthcare analytics, this involves having the core administrative systems working properly such that basic reporting requirements can be met, including the ramp up for many compliance demands resulting from the ACA. Often, and with the increased regulatory demands and competitive threats placed on healthcare providers and payers, this involves a review of legacy systems and a total rethinking of IT capability and functionality. And this can be very expensive and take years to complete. So, for most organizations, there is a commingling of the old with the emerging new information capability, but during the process there might be scant results that demonstrate the worth of analytics to improve the business.

The real worth of analytics comes into play when the organization has a well-functioning integrated information management system. For many of the industry challenges, this requires an integrated view of the patient-member-person in order to address behavior change for prevention and treatment and to maximize marketing offerings that are tailored to the individual customer.

Ultimately, the self-actualized, analytics-optimized organization will be able to use its real-time, integrated, comprehensive information to address new market opportunities and indeed to master continual transformations of the organization.

However, this need not be a linear progression and the perfect should not be the enemy of the good. Value needs to be extracted from health analytics all the time. It needs to show a clear contribution to earnings. It cannot wait for the perfect system to be erected because during the building process new demands will change what perfection looks like. And if the analytics is not clearly contributing to the bottom line, it will be considered a cost center and will be subject to budget reductions when times get tough for the business.

Top-Ten List of Healthcare Analytics Challenges

Figure 2.3 lists the most important analytics challenges, and the sections that follow discuss each challenge.

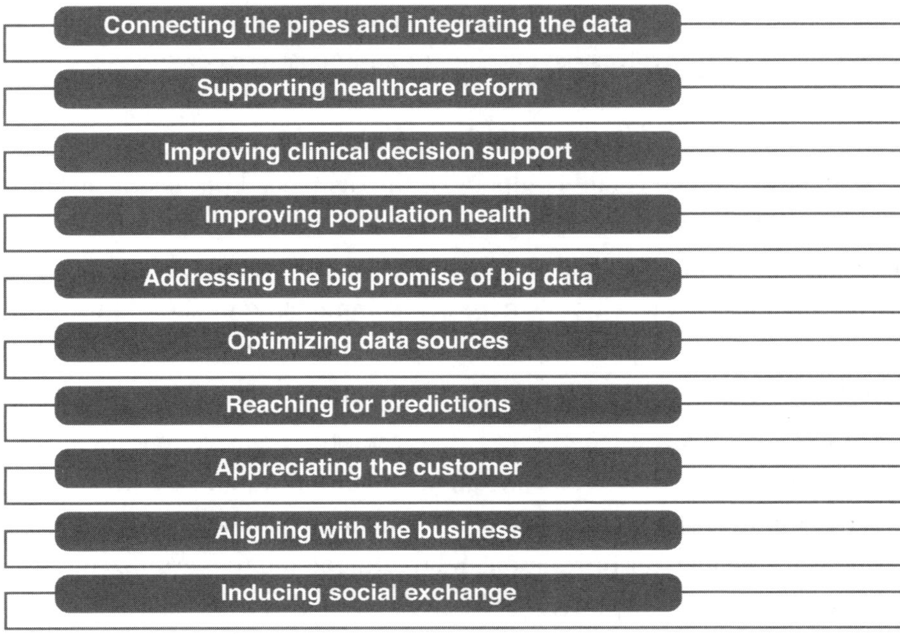

Figure 2.3 Top-ten list of healthcare analytics challenges.

Connecting the "Pipes" and Integrating the Data

This is the most critical challenge to the full expression of analytics in healthcare. In provider settings, it starts with the electronic health record. Digitizing the medical record has been discussed for decades and proceeded in fits and starts until 2009, when progress accelerated because of the federal Health Information Technology for Economic and Clinical Health (HITECH) Act, which was to advance the use of health information technology. HITECH, through Medicare and Medicaid, provides incentives to physicians and hospitals that adopt and demonstrate "meaningful use" (MU) of EHR systems. According to a 2012 National Center for Health Statistics (NCHS) Data Brief,[37] more than 50% of all physicians had adopted an EHR system by the end of 2011 and, of the remaining 50%, half plan to purchase or use one already purchased within the next year. Similarly, a 2012 survey of U.S. hospitals indicated that EHR adoption increased from 16% in 2009 to 35% in 2011.[38]

This is great progress, but there is a long way to go. These systems facilitate communication among doctors and caregivers and various departments within hospitals and associated outpatient physician practices for the delivery of clinical care. However, connecting to providers outside the hospital walls or to other healthcare systems and then to other health systems in the community is a challenge because of "dueling" systems. That's the purpose of the health information exchange (HIE), to mobilize healthcare information across organizations. But for now, if a patient goes from one hospital system to another, the mode of medical record transport is usually the patient carrying the record, X-rays, and lab findings. And understanding whether the patient received services from other health providers, such as public health agencies or schools that provide vaccinations and tests, is infrequently captured in the medical record.

Analytics staff within hospitals are largely concentrating on implementing the EMR and getting people communicating better and improving care in the process. But after the data collected in these systems is used for that specific clinical purpose, most of it becomes "digital waste" and is not reused for other analytic purposes. It is stored away, like the cryopreservation (freezing) of dead human beings awaiting the technology to bring them back to life. The difficulty in using the EMR data is that most of it is unstructured. There are ways to unlock unstructured data and to transform it into standardized data that can be reliably retrieved and combined with like data from other systems. However, this is still a work in progress and best-practice illustrations of successful implementations are not at all abundant.

Integrating data from across the enterprise and beyond is also a major construction undertaking. This involves the building of sophisticated data warehouses that bring together a wide scope of information including clinical, financial, administrative, and more. A good example of this is from the University of Pittsburgh Medical Center (UPMC). It recently announced a five-year, $100 million investment in enterprise analytics, in partnership with a variety of technology partners including Oracle, IBM, Informatics, and dbMotion. It hopes to address big clinical and genomic prediction questions such as "What if a doctor could easily predict which treatment would be most effective and least toxic for an individual breast cancer patient, based on her genetic and clinical information?"[39]

There is great hope and promise to get clinical processes and the healthcare system "wired." Organizations like UPMC are willing to make very significant investments over a long-term horizon to achieve potentially transforming goals. When these early innovators achieve success, others will follow. Whether the "pipes" goal is to digitize the medical records of a hospital or health system, or to integrate diverse data into data warehouses, what is clear is that the digital construction process is where the majority of analytics time is spent. The challenge will be that while huge investments are made in improving IT systems, the business is not deriving much value from it today to improve business results.

Supporting Healthcare Reform

After 50 years of debating what healthcare reform should be, it's finally done (mostly). The policy analyses and politics are mostly over and now the hard work of implementation is underway. Analytics will be central to at least three key areas of the legislation, including insurance reform (especially health insurance exchanges), CMS innovations (especially ACOs and the consumer operated and oriented plans [CO-Ops]), and health information technology (HIT) (especially meaningful use).

The health insurance exchanges will open up the "insurance malls" in October 2013. Although most states, including almost all of those with a republican governor, have decided not to form their own exchanges and will probably defer to the federal government to organize and run it, others are well on their way to being up and running on schedule. The exchanges themselves are a huge HIT project involving the assessment of eligibility, enrolling individuals into coverage, and the collection of insurance premiums. Additionally, the exchanges will bring together quality, cost, and benefit information and present it to applicants to facilitate their choice of plans.

The larger analytic challenge is with the health plans as they move to a direct-to-customer business model. This will require a great deal more knowledge about their new customers in order to facilitate member acquisition and member behavior change to improve outcomes and reduce costs (now that they cannot make profits solely on skimming the healthiest people). This will involve a greater understanding of marketing analytics, including lots more data about individuals' characteristics and risks and about how they have used the various components of the healthcare system and the insurance system. This will require a 360-degree view of their customers and the underlying analytics and IT infrastructure to support it.

Some payers are also changing the nature of their business. They are supporting ACOs by becoming data aggregators to help them assess risk, utilization, outcomes, and shared savings through a better understanding of claims, member, and other data.

An unintended, positive outcome from reform is the coordination and communication among analytics staff within organizations. These staff from finance, quality improvement, insurance, and various business units have a new organizational imperative. And because lots of money is at stake, whether it be for meaningful use, readmissions, or patient satisfaction, a "burning platform" to address the money concerns has driven people together like nothing else before.

Improving Clinical Practice through Decision Support

Connecting the pipes and integrating the data are critical, foundational, developmental stages for analytics. Making use of the data and turning it into useful, accessible, timely, and user-focused information for decision making is where the payoff happens.

Improving clinical decisions encompasses a number of initiatives:

- Shaping care through decision rules. These include rules for care protocols, drug interactions, diagnosis, and order sets that can be included in EMRs.
- Monitoring and optimizing performance through balanced scorecards and dashboards that are used for management review and interventions.
- Supporting physicians and caregivers with tools for clinical decision making at the point of care.

Analytics supports the latter two initiatives the most, as described in the following sections.

Dashboard

Dashboard reporting is common in healthcare. Balanced scorecards provide a critical lens on the performance of the key success factors of various domains that encompass a complete picture of the business. For hospitals, the domains usually include quality of care, patient safety, patient experience, access, and financial outcomes. Many indicators are rolled up into a summary, composite measure for each domain. Targets are set for each indicator and domain composite. The dashboard tracks progress relative to target, usually on a monthly or quarterly basis. Visualization of progress toward goals includes the use of traffic stoplight signals of green, yellow, or red to indicate, at a glance, whether management needs to intervene. The component indicators of each domain can be accessed with a click on the domain composite. The reporting is meant to be simple and direct and home in on areas in need of attention.

It is important to note that these are management reports and are largely "top down" intervention tools. For example, if performance targets on a given domain are not achieved, managers are held accountable and need to explain the causes and how remedies will be applied. It is also important to note that these reports are at an organizational level and seldom go down to the physician level, although the data should be available at that level for physician-related indicators such as quality and patient safety.

There are a few rationales as to why dashboards do not go to the granular level of the physician. Most notably, practicing physicians might not view this aggregate information on their patients as valuable to them in the treatment of individual patients. Also, publicly reported performance information on individual physicians (or teachers, for that matter) incurs the wrath of those who are evaluated because they feel that the metrics are not scientific, risk adjusted, or fair.

In the Chapter 6, "Sports," it will be clear that healthcare can learn a lot from sports, believe it or not. Lessons can be learned about measuring individual performance and why a concentration on the individual rather than the team can be counterproductive for both industries.

Supporting Physicians

Physicians need just-in-time information to support them in the diagnosis and treatment of individual patients. Real change in healthcare takes place "on the ground" at the physician/patient level. But the context of the clinical encounter is demanding on a number of fronts. One is that the practice of medicine is very fast-paced because of the efficiency demands placed on physicians with appointments that last less than 10 minutes. Hence, analytics needs to demonstrate its value to provide information, just in (real) time, and customized to the needs of the user.

However, physicians do not use information optimally for a variety of reasons. The first reason is that it is impossible to keep track of all the emerging research and related insights about diseases and new treatments. Doctors just do not have enough time to digest this information in their scant leisure time. The second is that it is not always possible for doctors to get the information they need on a given patient because they might not be able to find it if the records have not been digitized or integrated. The third reason is that even if this information is available, it might not be used because physicians are trained in and have a strong preference for intuitive thinking. They do not have the inclination to sort through lots of information and decision maps to make data-driven decisions. One indicator of the need for more data-driven decision making in medicine is that diagnostic errors occur about 20% of the time.[40]

One emerging solution that might transform the practice of medicine is machine learning. IBM demonstrated a compelling use of machine learning (and natural language processing and predictive analytics) with its Watson technology by beating two grand champions on the *Jeopardy!* TV quiz show. IBM is now moving beyond quiz shows and working on healthcare solutions, mostly in the area of differential diagnosis. One of the institutions it has partnered with is Memorial Sloan-Kettering Cancer Center.[41] The goal according to Sloan-Kettering is to have the technology gather and assimilate information from the research literature and from the Center's vast clinical experience documented in its medical records and other files to "bring up-to-date knowledge to the bedside of every cancer patient."[42] Watson might be able to do this through its capabilities to read and understand language, interact with humans, remember everything, and provide answers to real-time questions. How the information will be delivered to the physician, how it might transform the practice of medicine, and whether physicians will embrace the technology are all very important, open questions. It is certainly a very bright star illuminating and guiding the emerging field of machine-enabled clinical decision support.

Other industries do not have a lot to teach healthcare in this area. There is certainly a lot of theory and research about machine language and its related technologies, but no clear evidence of its utility to improve business results. Healthcare is forging the way and other industries might be able to learn from it.

Improving Population Health

Population health is more a slogan than a business strategy at this point and might wallow in the background until a few things happen. Payment policy needs to change from fee-for-service to global payments; outcomes need to be measured and adequately rewarded financially; ACOs and medical homes need to move beyond a focus on coordinated care for sickness and embrace prevention just as thoroughly. As discussed earlier, many of the shortfalls and the opportunities for improvement for the health system stem from less than full attention to population health issues. *This is perhaps the biggest area where healthcare can learn from other industries and is addressed in almost every industry comparison.* It might seem counterintuitive that other industries take a "population" perspective. Banking, retail, and political campaigns market and attempt to sell and serve everyone who could possibly be a customer, unlike healthcare, which concentrates on the sick when they seek care and leaves the majority of members and patients relatively unnoticed. This is the group that needs to be focused on for prevention and early intervention of diseases.

Addressing the Big Promise of Big Data

There have been big claims about the potential for "big data" in healthcare. McKinsey states that "if U.S. healthcare were to use big data creatively and effectively to drive efficiency and quality, the sector could create more than $300 billion in value every year."[43] And technology vendors such as SAP state that "with the right information at the right time, anything is possible...and with real-time, predictive analysis comes a shift toward an increasingly proactive model for managing global healthcare."[44] Big data is sometimes subsumed under personalized medicine with a strong emphasis on the integration of genomic and phenotypic data. The result of such an integration could lead to answers to important questions, such as whether a certain chromosomal variation is related to a disease, which could then fuel predictive and preventive medicine in the future. Moreover, boundless data in the form of personal and public and integrated medical data, in combination with the right channel—especially mobile—enable much better identification of high-risk patients, more effective interventions mapped to their specific needs, and closer monitoring.

Big data is also big business. In addition to the traditional technology players such as IBM, Oracle, SAS, and many others, more than 200 new businesses are developing healthcare solutions.[45] The focus of these innovative companies is less on data management or traditional retrospective reporting and more on predictive modeling, direct intervention for behavior change, and using mobile apps that track location, daily activity, and patient-reported events.

Healthcare lags other industries, including banking and retail, in its use of big data because of the challenges with complex and unstructured data, the reluctance to use external data, data integration issues, and serious concerns about patient confidentiality. More important, perhaps, is the need to ask the right questions and to formulate the use cases that make sense for the business. Finally, achieving the big potential of big data is a way off. First movers may realize breakthroughs to outpace the competition if they have the stomach for the risk and cost involved. But, there are many ways to chip away at big data, especially by using external data and omnichannel platforms that can address business needs today. A few of these are the direct conclusions of this book.

Optimizing Data Sources

The world is full of relevant data and a lot of it resides outside of healthcare. External data can address very specific questions about healthcare business issues to change people's behavior, ranging from marketing to early detection of diseases. And it can do it cheaper and better—and do it now—while the big 360-degree integration project is being perfected and the big data neural transformation of the world evolves.

These data come from privately aggregated and publicly available databases on a wide range of personal attributes and can focus on specific interventions to improve health. For example, data on height and weight can be used to calculate a body mass index (BMI) that can indicate the degree of obesity, which is predictive of diabetes onset. Given the enormity of the diabetes problem in the U.S., early identification is very important to save lives and dollars. Yet the basic information on height and weight is not collected on the usual claims forms or member registration forms. Although it can be collected in EHRs, it might not be extractable from them for population health purposes. Additionally, it comes from patients in treatment and overlooks those people who have not yet entered the healthcare system. Healthcare has attempted to gather the data directly from people on health risk assessments (HRAs), but the response rate is too low for the data to be useful. And the people who tend to fill out the HRAs are the very healthy people because documenting good health is often linked to monetary incentives for wellness programs.

Height and weight are at the tip of a very deep iceberg on what is available about individuals in this über-surveillance society. Later chapters document how other industries have teamed with "data snatcher" companies that collect reams of data on people from public records, purchases, Internet activity, polls, and more. Then, the newly mined data are used to literally win the presidential election, predict life changes, and amass patterns indicative of terrorism. Although there are many privacy and ethical concerns about these data, the use of it is rampant and the business successes have been noteworthy.

The healthcare industry has been reluctant to use these resources when compared to other industries such as retail and banking. There are several explanations. Lawyers get squeamish about the collection and integration of *any* data that has an ID on it. HIPPA *might* be a barrier. There is a knee-jerk reaction that it might be all right for other (lesser) industries, but not for healthcare. And there is enough unused healthcare industry data to keep IT busy for decades.

A case is made in the industry chapters that healthcare is blinded to the potential usefulness of external data to its detriment. There are many business purposes ranging from customer analytics for marketing, to predictive analytics to identify people at risk, to behavior change analytics to improve engagement in one's own health improvement.

The technology is available to collect, integrate, and make all sorts of computations on the data. One problem is whether there is a compelling ROI in doing it all. This revolves around the important issue of what the business purpose is and whether the business can execute on the data. For example, the data on BMI can be obtained at a cost of about $20 per 1,000 matches to a health plan's membership list. This would seem to be a modest cost even if the legal fees to get it were to add an additional overhead cost of $5 per match! Integrating the data is not difficult and stratifying people by obesity is relatively straightforward. This seems like a worthwhile investment so far.

According to the American Diabetes Association, there are 79 million people in the U.S. with pre-diabetes and 7 million with undiagnosed diabetes. The medical expenditure for people with diagnosed diabetes is 2.3 times that of people without diabetes.[46] So the potential cost savings from early intervention to thwart diabetes is mammoth. And the analytics investment is low.

The concern is whether the health plan has the behavioral change technology to actually achieve change that would result in obesity reduction. The first step, and it is an important, fundamental contribution of analytics, is to identify. But a whole series of implementation linkages needs to unfold for the analytics investment to pay off. And this requires that the business is coordinated, integrated, and impactful.

Reaching for Predictions

Predictive modeling has been around a long time. The advances that have occurred to make this a powerful tool are not so much the statistics or theoretical models but the technology advances to collect, capture, integrate, process, and make computations on large amounts of data. Virtually every valuable analytic solution in healthcare and in other industries involves predictive modeling. Describing what happened yesterday in static, standard reports is so, er, yesterday. The breakthrough with analytics comes with predicting what will happen tomorrow.

Other industries described in this book include many examples of how modeling on large, multifaceted databases produces important business results. Retail, credit, banking, and presidential campaigns can predict individual behavior before the individual is aware of it. For example, credit companies have determined the major risk factors related to default, and one of the most predictive is divorce. They have created divorce algorithms based on individual purchasing records that indicate high risk of divorce. Retail can predict what you will buy before you know it. And on and on.

Healthcare has a great need to predict risk and outcomes for specific segments of patients/people/members. These potential predictive modeling sweet spots from other industries and the potential adaptability for healthcare are a major theme of the book.

Appreciating the Customer

People are called many names in healthcare, including patient, consumer, member, citizen, and client...but hardly ever customer. This blind spot in healthcare to recognizing the value of the individual as the customer is both understandable and delimiting.

Previously, we noted that customers are defined as the ones who pay the bills, and the individual is a minor player in that regard, superseded by employers, insurance companies, and provider groups. So thinking of the patient/member as a customer can be a counterintuitive. And the term "customer" might have a negative connotation to some in healthcare. A customer might be regarded as someone you have to "deal with" who needs to be "converted" for the sale. And the track record of some business functions such as marketing and customer service demonstrates that what is done *to* customers can be annoying and intrusive or downright antagonizing and profiteering.

The appreciation of customer value must go beyond "the sale." It needs to be about a relationship over time. In the absence of a relationship and some element of loyalty, the customer can shift companies at the click of a mouse to get the lowest price or the flashiest coupon offer. Mahatma Gandhi thinks of customers in a respectful way:

> *A Customer is the most important Visitor on our premises.*
> *He is not dependent on us, We are dependent on him.*
> *He is not an interruption of our work, He is the purpose of it.*
> *He is not an outsider to our business, He is a part of it.*
> *We are not doing him a favor by serving him, He is doing us a favor by giving us an opportunity to do so.*

Healthcare needs to be more patient-centric and more customer-centric to achieve the important goal of helping the individual change behavior. The absence of appreciable "people engagement" in healthcare has long been lamented, but progress in areas such as patient-centricity, shared decision making, prevention, medication compliance, and having more "skin in the game" has been slow and exerts a significant drag on outcomes and costs.

There are indications that this is changing for the better in healthcare. The ACA has two important provisions that will require a focus on the customer. These include the health insurance exchanges, which will require a retail relationship with individuals. Additionally, CMS value-based performance payments to hospitals reward hospitals with good patient experience ratings based on patient ratings on their hospital stay. Also, the patient engagement movement has worked hard to change the behavior of healthcare providers to include more involvement of patients and families to improve health outcomes.

Jesse Gruman, president of the Center for Advancing Health, states it eloquently: "If we don't connect with at least one trusted clinician, show up when we need to, get the tests we agree on, use effectively the drugs and devices you recommend, carefully follow directions after a hospital stay and try our damndest to lose the weight and walk around the block more often, medical interventions are squandered. Our suffering continues. Time and money are wasted: yours and ours."[47]

Healthcare can learn lessons about customers from all the other industries in this book. This includes having a full view of the customer through lots of data, using predictive analytics, and implementing mass customization to capture a market of one.

Aligning with the Business

Analytics can do many wondrous things but without the support of the business it can achieve very little. After all, the business units fund analytics. Analytics exists to support the business unit needs. Similarly, technological solutions are relatively easier to develop than are the sociological steps needed to absorb the technology into successful management actions.

Throughout this book the discovery of the best analytics starts with an understanding of the industry's purposes, challenges, core functions, and unique business sweet spots. The best analytics emerge from this admixture.

In healthcare much of the analytics work is currently done away from the business, and for all intents and purposes it could be in a different country. It has to do with building the pipes mostly within the IT organization. Business agrees with the building of a data warehouse and says, "Let us know when it is done." But when the inevitable implementation snags occur and IT asks for more funds to complete the project, the business can and does balk. The business demands value generation every day and this has to include analytics as well.

When analytics works hand in glove with the business, amazing things can happen. The analytics function grows to its full potential and the business succeeds in ways that were impossible without the partnership with analytics.

Inducing Social Exchange

Mobile technologies are creating a social revolution. These technologies allow the full expression of the person's long-held desire for convenience, simplicity, immediacy, autonomy, and technology that actually works for them on a 24/7 basis, and is a pleasure to use. Mobile is the communications nexus for people. We are continuously affixed to the world through it. And most important, we control the flow of the information.

Mobile technologies have the potential to engage and empower individuals to improve their health status. The applications are amazing and simple, at least to the user. Mobile applications include a GPS locator, camera/camcorder, and the capacity to send and receive digital files, including pictures and videos and biometric data. This combination of capabilities can allow a person to photograph a mole on their skin and send it off for immediate analysis on whether it might be melanoma. With simple attachments the smartphone can also facilitate the photographing of a child's ear to get more information on the pain; or track movement to detect anomalies from usual patterns to predict mental illness relapse; or analyze one's breath for unsafe levels of blood alcohol.

Mobile communicators also leave a digital trail of their clicks, along with data about their friends, purchases, ratings, and more. So in addition to the mobile phone being a communications nexus, it is also an individualized data repository.

The challenge and opportunity for analytics is to facilitate the proliferation of data generation and to tap into it. Tapping into it in a covert way through surveillance of data streams has clear but limited utility. The real potential lies in gathering much more relevant data from individuals with their consent and engendering their partnership to engage in data-sharing activities that help them improve their life. There is much more to be learned from a person's head than from their data streams. Political campaigns provide the best lessons on engaging, empowering, and learning from customers.

3

Retail

More than any other industry, retail has developed leading-edge analytic techniques to capture the minds and wallets of customers.

Introduction

The fundamentals of the retail industry are straightforward: Buy or make goods, place them on a shelf, and sell them at a profit. There are many types of stores that sell merchandise, including department stores like Nordstrom and Saks, large chains like Walmart and Target, warehouses like Costco and Sam's Club, and Internet sites like Amazon.

The industry is huge and 50% larger than healthcare. Retail sales in the U.S. were about $3.8 trillion in 2010, according to the U.S. Census Bureau.[1] This represented about 27% of the economy in terms of GDP. It employs more than 14 million people in the U.S. alone.[2] In comparison, the U.S. healthcare industry had revenues of $2.6 trillion in 2010, represented 18% of GDP, and employed almost 12 million people.[3] Walmart is the world's largest retailer. In the U.S. it had 3,868 stores and $264 billion in net sales in 2011.[4] The largest U.S. healthcare player in 2011 was United Healthcare Group (an insurer), at $102 billion in revenues.[5]

In terms of the current economic landscape for the industry, it has recovered modestly from the lows hit during the Great Recession of 2008. However, poor holiday sales in 2012 might portend a reversal in the upward trend.[6] Industry watchers characterize the future as being "hypercompetitive," both online and in stores, necessitating the mastery of many tough challenges, not the least of which is being much more data driven.[7]

No other industry is so customer focused. The art and science of marketing has been fully developed in retail. But the times are changing. Customers are becoming more demanding and less loyal. And power has shifted to them. Technological innovations enable full transparency on products, pricing, peer satisfaction, and availability. Consumers can touch and feel merchandise in the store, use their mobile device to instantaneously compare alternative prices, and buy from the lowest-cost provider either online or offline.

Retailers' approach to customers has also changed dramatically, driven primarily by the squeeze in profits caused by a number of factors to be discussed shortly. In the "old days," shopkeepers had a one-on-one relationship with customers that offered the opportunity for deep product understanding, reliable customer experience, and the sustenance of trust. Today, most retailers have reversed this and have changed to a decidedly less personal, mass-retailing approach. Few stores, except those at the high end, are likely to emphasize personal customer relationships. But customer analytics can bridge this divide somewhat and offers the technologies to provide more high-tech personalized and tailored services with scale and efficiency while avoiding the seemingly unaffordable costs of high touch.

Retailers are at the cutting edge of using analytics for marketing. A treasure-trove of new data is available through records of purchases online and off and through shoppers' clicks online and apps on mobile devices. Additionally, retail is at the cutting, and potentially bleeding, edge of exploiting a wide range of personal data to increase the effectiveness of marketing and sales tactics. This newfound intelligence promises a significantly higher return on marketing investments. However, there is an emerging backlash against using personal data without consent, or losing it, or selling it, which could lead to serious injury to the brand.

Industry Challenges

Industry challenges are important determinants in defining and making—or breaking—a business. For retailers, there are three superordinate challenges: the economy, the battle between bricks and clicks, and winning over customers.

It's the Economy

The most serious exogenous factor affecting the retail industry, like every other industry, is the economy. Spending for retail goods is driven by personal income, consumer confidence, job growth, and interest rates, all of which have suffered during the Great Recession. Profits in the industry are razor thin compared to other industries, averaging between 2% and 4%. Profits were particularly low during the recession, plunging to 1.4% in 2008 and 2.4% in 2009 and returning to the relatively low prerecession level of 3% in 2010.[8] Another pressure on profits related to the economy has been the increase in (overall) commodity prices which increased more than 100% during the period for many items including food, fuel, and agricultural raw materials.[9] These combined economic pressures threatened the survival of retailers and many went bankrupt, including Sharper Image, Linens 'n Things, Bombay Co., and the mail-order firm Lillian Vernon. And, as with all industries under cost pressure, there has been considerable consolidation, for example, the acquisition of Sears by Kmart and the merger of May Department Stores with Federated Stores.

The Battle Among and Between Bricks and Clicks

Superstores are battling with each other on every major block of commerce. Pricing is very competitive. The in-store experience has to be compelling. And to add more pressure to the mix, online retailers are gaining ground with substantial year-over-year growth in sales.

Brick-and-mortar stores are very expensive to operate relative to online sites. So the lure of brick-and-mortar has to be more than a simple place to view and buy merchandise. It needs to be unique, even a destination experience:

- Malls like the Mall of America are so large and have so much entertainment wrapped around the stores that one needs a travel agent to navigate them.
- Some stores are more like wired living rooms that happen to sell coffee, like Starbucks.
- Others are hip and trendy and cater to younger segments, like H&M and Urban Outfitters.

- Still others are like four-star hotels, beautifully decorated and with piano music in the foyer, like Saks and Nordstrom.

However, for an increasingly large segment that goes to warehouse stores, including the unvarnished concrete floors, fluorescent lighting, and industrial shelving, price reigns supreme and people are willing to forgo pleasant surroundings and accept nary a customer service person in site.

The clicks are gaining on the bricks. It is projected that online sales will amount to about $250 billion by 2015, or about 8% of all retail sales.[10] The projected compound annual growth rate for online sales is 10%. A majority of online shoppers think the best deals are available online.[11] As shoppers get more comfortable with paying online, online delivery times improve, and the types of purchases expand from relatively small-ticket items to a broader mix, the growth potential might get much larger.

Offline stores are not going away. Major retailers sell the vast majority of merchandise in-store rather than on their own Web sites. And different customer segments have different preferences in relation to bricks and clicks. But the overarching trend toward impersonal selling, with price being the utmost buying factor, makes it harder for the old bricks formula to work, which included good customer experience and brand loyalty. As stated earlier, analytics might be an answer to approaching a market of one through radical personalization based on innovative uses of big data.

Winning Over Customers

Marketing is all about customers—acquiring, retaining, growing, and optimizing them. And treating customers well is good for the bottom line. A recent multi-industry survey found that 64% of consumers switched from at least one service provider in the past 12 months—for example, a bank, utility, wireless carrier—due to poor customer service.[12] They conclude, "As consumerism trends continue, we expect customer service performance to emerge as a source of differentiation and contributing factor to loyalty and retention." But as traditional customer "amenities" wane, there is a paradox on what constitutes a good and profitable customer experience.

People definitely have a preference about the relationships they want to have with companies, and many are loyal and active advocates on a company's behalf as a result of good customer experience. We all know companies we enjoy interacting with and others that are a pain to deal with and are best avoided.

Customer experience can be measured. Forrester Research rates customer experiences across 14 industries, including airlines, banks, credit card providers, health plans, hotels, insurance firms, Internet service providers, investment firms, parcel shipping services, PC manufacturers, retailers, TV service providers, utilities, and wireless carriers. It defines customer experience in terms of (1) meeting needs, (2) being easy to work with, and (3) enjoyability. It produces a ratings and rankings by industry and across the 133 companies in the survey.[13]

The industry with the highest average score is retail, with a score of 82 out of 100. Following closely behind is the hotel industry at 80, and parcel delivery/shipping firms at 78. Retailers take 12 of the top 20 rankings. The highest-rated retailers on the top-ten list were a mix of store types and included Barnes and Noble (#1), Amazon (#4), Kohl's (#6), JCPenney (#7), Macy's (#7), BJ's Wholesale Club (#7), and Costco (#7).* Note that all of these retailers, with the exception of Amazon, have online and offline outlets. However, Amazon had $48 billion in online sales, which was more than six times the online sales of all these other stores combined.[14] And offline sales were 30 times greater than online sales for these non-Amazon stores.[15]

In comparison, health plans/systems are dead last among all industries, with an average score of 51. This score is even lower than those of credit card providers and Internet service providers. Kaiser Permanente is the highest-rated health plan/system of those sampled, with a ranking of 75 out of the 133 companies rated. Medicare is 102, and Anthem, Medicaid, and United Healthcare are close to the bottom of the list at 129, 130, and 131, respectively. In between the top-ten-rated customer experience companies and the highest-rated healthcare company (Kaiser), there are many brands representing many industries, including SunTrust

* Other categories of the retail industry not included in these sales figures are gasoline, food service, and automobiles. Note that the survey includes a small number of retailers and appears to include the largest ones. The destination retailers mentioned earlier were not in the sample.

Bank (14), Federal Express (27), Southwest Airlines (28), Charles Schwab (40), MasterCard (57), and Cablevision (71).

The curious thing is that the essence of a healthcare company is to make people feel and function better; thus, it would seem it would have a greater opportunity to satisfy customers (if they did their job well) than a credit card company, a bank, or an investment firm, which handles monetary transactions. Presuming that the data are reasonably valid, it would appear that something has gone awry with customers in healthcare, especially with health plans. (Note that other types of insurers, for example, for cars and casualty, score much higher than health insurers, at 71% on average.) Given the low performance for the healthcare industry, the opportunity for improvement is huge. For example, if the average health insurance plan performed like the average retailer, there would be a 61% improvement.

Industry Strengths

Industries develop unique responses to challenges and market demands, which result in strengths that can be optimized for business success. For retailers, there are three areas of industry strengths including optimizing customer value, customer analytics, and mobile devices and social media.

Optimizing Customer Value

The business success of retailers depends on optimizing the customer's perception of value. There are six traditional dimensions of the value equation, including store factors, service factors, merchandise, price, supply chain, and technology:[16]

1. Store factors have to do with the look and feel and overall pleasantness of the store experience. Some stores emphasize beautifully decorated atmospherics like Crate and Barrel. Some emphasize sparsely appointed and industrialized surroundings to highlight efficiency and low costs.
2. Service factors are primarily about how the customer is helped in shopping, purchasing, and returning merchandise. Some stores are surprisingly stingy with customer service. Others place a lot of emphasis on servicing the customer, including Trader Joe's, the Container Store, and Lowe's.

3. Merchandise is the raison d'être for retailers and is why customers come to the store or the Web site. The store needs to carry what the customer wants and do it in a reliable way; for example, having the item on the shelf when the customer comes calling.
4. Price is a critical factor, especially when the store is selling (most) items that are also available from competitors. Pricing is dynamic and can change by the minute on Internet sites depending on supply, demand, and competitor pricing, or it can be dependably stable and low as with some warehouse stores.
5. Supply chain involves the efficient and effective integration of suppliers, manufacturers, warehouse, stores, Internet sites, and transportation such that merchandise is produced and distributed in the right volumes, for the right locations, and at the right times.
6. Technology is an important differentiator for retailers as it is for most industries. In the next two sections we highlight the importance of customer analytics and mobile devices and associated applications as particularly important technology solutions. Other important technology approaches are used in supply-side management. For example, CPFR (Collaborative Planning, Forecasting, and Replenishment) systems use sales data to inform the need for product replenishment. RFID (radio frequency identification) is the use of chips placed on merchandise to track its location through the supply chain. And CRM (customer relationship management) systems collect lots of data from customers, including purchases and customer service encounters, with the goal of building good relationships and loyalty with high-value customers.

Customer Analytics

> *"Winning retailers will have a better understanding of their customers than their competition."*
>
> —"Retailing 2020: Winning in a Polarized World," PricewaterhouseCoopers/Kantar Retail, 2012

Knowing the customer is a function of marketing and the retail industry does it the best. It uses marketing analytics extensively to this end. Marketing is all about understanding, predicting, and influencing customer behaviors in relation to buying and loyalty. It does this through four stages:

1. Acquiring new customers through marketing campaigns that target specific subpopulations of consumers (segments) with tailored messages and preferred media.
2. Keeping the customers you have through ongoing personalized communications and better-informed customer service.
3. Growing revenues through cross-selling associated existing products and developing new ones.
4. Optimizing return on investment (ROI) of marketing campaigns by concentrating on customers with the highest customer lifetime value.

The conversion rate, that is, the percentage of people who shop for a particular item and then buy it, is an important performance metric of retail marketers. These rates can range from a low of 0.2% to a possible high of 38% depending on the method.[17] The marketing methods and associated conversion rates* are listed here:

- Information cost reduction: 0.2%–2.9%

  Target addressable mail to reduce postal costs and lower cost of communications.

- Fact-based marketing: 1.9%–4.8%

  Identify loyal customers, target specific customer segments, use analytics in campaign management.

- Information sharing: 6.2%–18.7%

  Integrate multichannel buying data; for example, browse in one channel, purchase in other, and pick up in another.

- Information responsiveness: 16.9%–38.2%

  Analyze "raw" data from streams of customers' social media commentary to address changing moods and factoring these into point-of-sale rules.

* These percentages are cumulative; e.g., fact-based marketing includes the gains achieved in the prior method.

Clearly, the more technically driven approaches are producing the highest conversion rates.

Are these conversion rates good? Certainly, it must be disappointing to achieve only a 5% success rate with fact-based marketing and to just double that with multichannel marketing. When comparing to another industry like healthcare, by one standard, it is pretty low and by another it is pretty high. If a patient has a problem (a need-to-buy in retail) and visits a doctor (the store), and is offered a treatment (a product or service), most patients will buy it on the spot (although an uncomfortably high percentage do not follow through with the treatment). On the other hand, health plans that attempt to change behaviors of its members, for example, to ask them to change from a brand-name drug to a generic or to fill out a health-risk-assessment survey, do worse than retailers.

The retail industry thinks that technology advances will bump up these rates. For example, a recent survey of retail chief marketing officers (CMOs) indicated that more than 80% say they expect to increase their use of four technologies in the next three to five years, including customer analytics, customer relationship marketing, social media, and mobile applications.[18]

Big data brings big promises and challenges to retail. More and more data are needed to understand customers better and better...and to do it faster than the competition. The hallmark of customer analytics in retail as compared to other industries is the availability of purchasing and loyalty data both online and offline. Additionally, retail is more adventurous in using personal data on customers than other industries. And huge volumes of data are available from customers' clicks as they browse through the Internet that are tracked by scores of firms that put cookies on consumers' computers.

One of the challenges is that retailers are struggling to figure out what all the new data means. The best is truly yet to come as analytics unveils the secrets of consumer behavior and uses predictive modeling and other techniques to improve customer relationships and buying behaviors. The trick will be to understand shoppers' passions and interests along with purchasing data and to use these data in real time to make the "next best offer." (Examples of these are in the next section.)

However, as retailers blaze into new frontiers in the use of big data, there are risks. For example, in addition to using purchasing and loyalty data, the industry buys publicly available data through data brokers on parameters such as the value of a house, religious preference, and more to improve its predictions. And it scrolls and scrapes personal data using cookies and software in mobile phone apps and the Internet. And it does this without the knowledge or consent of the customers it is tracking.

It also pushes the envelope in matching personal data with real-time interventions. For example, some retailers use cameras in digital advertisement displays and change the message based on photographable personal characteristics. If a passerby appears to be young and female, certain apparel promotions might appear on the display. Other retailers use cameras for face recognition, integrate these data with customer profiles, and then produce real-time messaging to store salespeople on approach and messaging to the client; for example, "Hello, Mr. Smith, are you interested in the latest sport coats from Ralph Lauren to wear on your next cruise?"

Many feel that the use of personal data without consent is getting "creepy" and leading to a bevy of privacy concerns. In fact, federal legislators are concerned with buyer classification scoring, which might redline certain buyers because they do not provide optimal lifetime values. This is analogous to credit ratings but is not regulated at all, unlike the credit industry.

Because it stores so much personal information, including credit card information, the retail industry is also prone to security breaches that can include hackers stealing credit card numbers and other sensitive information that can lead to identity theft and credit card fraud. These activities can tarnish the brand. But the real challenge ahead might be consumers' rebuff of the use of personal data, obtained through cookies and other covert tactics, for marketing. This might stop mass customization in its tracks and turn back the rising tide for online sales. More likely, retailers will evolve new approaches, including creating data-free zones, third-party reputation protectors, and new data-sharing relationships between seller and buyer that are transparent and provide mutual value.

Mobile Devices and Social Media

More and more consumers are perpetually connected to smartphones, tablets, and other mobile devices to search for products, find coupons, find the best deals, and make a purchase. For example, there are price-comparison mobile phone apps that enable shoppers to scan the bar code of products in-store and get a real-time comparison of prices from competing retailers.

Additionally, customers expect to communicate with retailers through "omni-channels" and use their preferred device at any given time, on a 24/7 basis. They expect that all the channels will talk with one another and provide consistent and reliable information about the store.

Smartphones have become a key tool for shopping. A recent survey by Deloitte indicates that 67% of smartphone users between 14 and 34 years of age have used their device to shop, and 55% said that smartphones influenced their decision to make a purchase. The survey also addresses the "influence" of smartphones on store sales in relation to product research and price comparison, which is different from using smartphones to actually complete the sales transaction. The study concludes that almost 20% of total store sales, or $689 billion, will be influenced by smartphones by 2016.[19]

Social media is just emerging as a tool for conducting e-commerce. According to Booz Allen Hamilton in a 2010 report, the market is embryonic with sales of $5 billion. It projects sixfold growth over five years and estimates that $30 billion will be spent by 2015. It will complement other channels, including stores and the Web. Social e-commerce includes platforms that sell merchandise like Groupon and Gilt; social apps that facilitate group buying, group-gifting, and social shopping; and additional functionality for present sites like Facebook and Twitter to enable e-commerce.

In addition to the opportunity for commerce, social media serves other functions for retailers. This includes driving awareness and "buzz" and generating leads. It also facilitates the improvement on conversion rates. And its unique type of communication in the social nexus improves loyalty and can foster advocacy of its product and brand.[20] An example is included in the following section.

Analytics Sweet Spots

Relative to the key business drivers just described, this section turns to selected analytic sweet spots that support business solutions that are unique to retail and offer promise for healthcare (see Figure 3.1). It boils down to four areas:

(1) predictions: who and when, (2) lifetime value, (3) market matching: who and how, and (4) mobile and social media: buzz and advocacy.

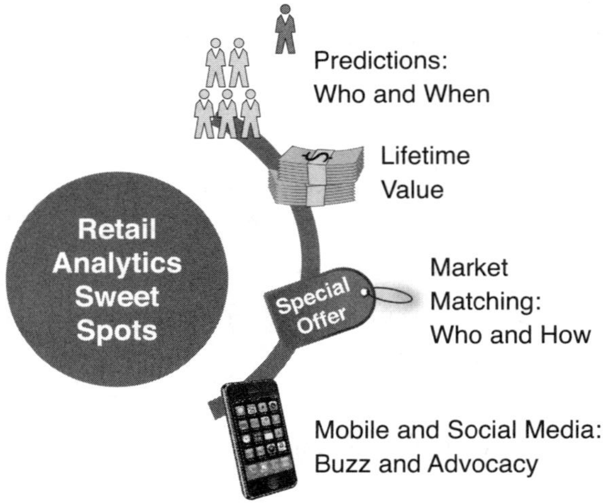

Figure 3.1 Retail analytics sweet spots.

Predictions: Who and When

> *"Just wait. We'll be sending you coupons for things you want before you even know you want them."*
>
> —Andrew Pole in "How Companies Learn Your Secrets," *New York Times*, February 16, 2012

The retailer Target uses predictive analytics to increase sales to pregnant woman.[21] Why pregnant women? They buy a lot of merchandise for the new baby and the new family lifestyle. Further, this particular life event presents a window of opportunity to influence buying habits. Most buying habits are fixed, relatively autonomic, and not amenable to new decisions rules. But "moments that matter" are the windows of opportunity that marketers home in on.

Target's prediction challenge was to identify the women through a "pregnancy prediction score" and to estimate the delivery date. The delivery date is critical because they determined that buying habits change during the course of the pregnancy; that is, different items are bought at different times. They did this by looking at what women on the store's baby registry were buying. It turned out that these women bought larger quantities of unscented lotion during the second trimester; supplements around 20 weeks; and soap, extra big bags of cotton balls, and washcloths when the delivery date was getting close. From these findings, they developed a model that provides a pregnancy prediction score and estimated delivery date based on purchasing data. Tens of thousands of women were identified.

The next challenge was to trigger new buying behavior by presenting coupons to the targeted women, for the right products, at the right time, via the right channel, and with the right message. Target will "be sending you coupons for things you want before you even know you want them." This was accomplished through the collection and integration of a wide array of data. Andrew Pole, a (former) statistician at Target, said, "If you use a credit card or a coupon, or fill out a survey, or mail in a refund, or call the customer help line, or open an e-mail we've sent you or visit our Web site, we'll record it and link it to your Guest ID...and we can buy data about you...[such as] charitable giving and cars you own." All these sources provide data on a customer's age, marital status and how many kids, neighborhood lived in, estimated salary, credit cards carried, Web sites visited, and more.

The results of the campaign showed that Target's Mom and Baby sales exploded. Overall revenue following this campaign increased dramatically. The campaign was presumed to be a great success.

So if this approach works for revenue enhancement for pregnant women, what about other marketing segments and "moments that matter," such as month three after a wedding engagement when the gift registry list is developed, or month two after the death of a parent when the inheritance is assimilated and thoughts about a vacation or investments are under consideration? These habit-changing moments are determinable, and sophisticated analytics can home in on the target segment with a high degree of accuracy.

However, there are risks to data mining of personal purchasing data. A father complained that his teenage daughter was receiving coupons for pregnancy products and thought it was inappropriate given that she was not pregnant. Turned out she was pregnant and the father apologized. But this is indicative of a widely held, uncomfortable concern about the covert gathering and analysis of personal data for the purpose of persuading/manipulating purchasing habits. On the other hand, some people appreciate it when trusted sellers make recommendations to them based on purchasing history; for example, Netflix and Amazon. It might even be a good trade-off relative to receiving reams of irrelevant advertisements for some consumers.

What are the implications for healthcare? First, of course, the healthcare and retail industries are very different in many respects. The retail industry sells merchandise used in everyday living, such as soap, clothing, and flat-screen TVs. Much of the merchandise sold includes such necessities as food and gas. Some merchandise brings happiness to buyers because it improves some aspect of their lives, such as a computer or a smartphone or jewelry. Retail items are also of relatively low cost. When it comes to healthcare, however, purchases can be very complex, involving high-tech treatments and extensive, ongoing contact with providers. And the treatments can be very expensive, requiring insurance to protect individuals from bankruptcy in the case of an accident or a terrible illness.

The industries are also very different in terms of commerce. People buy things in healthcare with "other people's money" because of insurance and are relatively oblivious to actual costs. Similarly, people tend to rely on others to make the buying decisions for them in healthcare, whether it is the choice of insurance plan made by an employer or the treatment prescribed by a physician.

The ways things are sold are also very different between the industries. Retailers have a different value proposition than healthcare, as discussed earlier. They sell through stores and the Internet and compete on price, quality, and innovation. Healthcare is sold largely in doctors' offices and in the hospitals they affiliate with. Transparency on price and quality in healthcare is far less complete than in retail.

The incentives given to consumers are also quite different. Retailers use discounting through coupons and rebates and other forms of price reductions, for example, the "blue-plate special." In healthcare, a monetary incentive is seldom given by the seller (physician) to the patient to get better. In fact, going to the doctor for a serious condition usually costs a lot out-of-pocket, the outcomes are not always known, and the experience is not often pleasant. And the only incentive from the doctor is to follow his expert advice to "do it or die." This should be sufficient, one would think, but then again the rate of medication compliance, that is, taking the medications as prescribed by a physician, is only about 50%.[22] Clearly, this "conversion rate" for patients to hear and act on the message to change behavior and improve one's own health is abysmal in healthcare and needs to change dramatically to achieve better treatment outcomes.

How customers are engaged is also different. For example, the clinical encounter is a very personal, one-on-one relationship and transaction. However, many patients might feel that their interactions with the healthcare system are very mechanical and depersonalized. As discussed earlier, most retailers have all but given up on the value of close personal relationships and have gone in the other direction, toward lower costs and mass customization.

So the real question is how might mass customization and the associated use of customer analytics, so refined in retail, benefit healthcare? Perhaps the answer lies in understanding the "patient" customer in different ways through the use of multiple streams of data, as used in retail. It might also lie in the relatively underdeveloped approach to improving population health rather than individual health.

In terms of predictive modeling, clinical decision making is quite advanced. It is certainly possible to predict the health outcomes of individual patients, based on their age, results from screening tests, clinical histories, genetic profiles, and so on. Indeed, that is the purpose of diagnostics. And making predictions on which treatments are the best fit and the estimated life years remaining (prognostics) are also core skills of care providers. The patient's data is applied to well-researched models of disease and the most likely fit is determined.

There are ways to improve the clinical encounter with retail-induced customer analytics, and examples are provided throughout the book. But the real lessons come into play with managing the health of populations. Although healthcare can be very good at treating illness, it is relatively weak in preventing illness in the first place and in partnering with patients to help them take the actions they need to engage in to improve care. The clinical data used in medical predictions is not sufficient to this purpose. Retail offers a look into the use of data outside the usual sandbox of an industry.

One example is the use of nonconventional, personal data to predict which people are obese and are at risk for serious health conditions, including diabetes and heart disease. The best indicator of obesity is the body mass index (BMI). It is very simple to calculate from data on height and weight. Yet, surprisingly, these data are not collected and stored by health plans. They might be available in electronic health records (EHRs) but these would include data only on patients who are presently sick. And the EHRs are usually not accessible by health plans. So how does a health plan get these important data on the vast majority of their members who do not seek healthcare, but might be at risk, and subsequently accrue more treatment and costs, if not identified early on in the progression of the disease?

Health plans might try to mine their claims data and come up with predictive algorithms that indicate the probability of obesity. But, again, the diagnostic codes on claims forms are for services provided for treated diseases, and most forms of obesity, with the exception of morbid obesity, are not diseases. It might be possible to look at the longitudinal records of patients who end up with diabetes and determine whether there are precursors that are reliable and valid indicators of diabetes and thereby develop a predictive model. (This is similar to what Target did in calculating the pregnancy prediction score.)

Somebody from outside the health insurance claims world might ask innocently and incredulously, "Why not just ask your members directly?" It's important information, they should care about their health, they will certainly provide it. That's what health-risk-assessment (HRA) surveys are about. Many companies provide incentives for employees to provide this information to a third party that can use it to motivate the individual to lose weight, stop smoking, drink less, and wear seat belts. Companies might pay $100 or so for employees to do this. But many employees might not want their employer even remotely knowing about their health risks, and therefore the only ones who might reliably fill it out are the healthy and young employees. In any event, the response rates to HRAs are low. But the response rate to surveys is better in other industries. Indeed, healthcare can learn from politics and its use of polling. Its customers get engaged because they think the cause is meaningful. In Chapter 4, "Politics (Presidential Election Campaigns)," we present some ideas about how to make the cause meaningful in healthcare.

Another approach is to take the blindfolds off and try a whole new way instead. Other industries buy reams of personal data and use it for commercial purposes to sell more products and services. In healthcare, the data can also be put to good use. In the case of height and weight, the information is obtainable from data purveyors along with a treasure trove of other data at relatively low cost (for example, $20 per 1,000 member matches). Almost all of these data are in the public domain and some come from more covert sources such as Internet clicks. This is much cheaper than $100 incentives for HRAs and more inclusive in terms of the response rate. The computational task is very simple—just two variables. The additional information obtained from the new personal file would also facilitate the building of many predictive models.

So why not try this whole new way to attack obesity and diabetes risk? We will discuss adaptation challenges in the last chapter. These include important considerations of privacy; for example, HIPPA rules, and organizational readiness. But what is clear is that the data and analytics are ready for prime time, at least to identify individuals. Getting the job done has more to do with the sociology than the technology.

What happens subsequently in the chain of events that has to take place to affect behavior change, in this case to lose weight? Identification in and of itself is quite useless. What is the model on the "moments that matter" to effect a habit change? Is sending out educational literature the answer? Would people respond to a health plan's coupon to join a health club, or would receiving an app to monitor food intake make a difference? Should health plans provide a Fitbit[23] monitor to track movement? Is the medical system the right system to lead the charge on the obesity initiative? Are the factors related to obesity more social than medical? Is the role of analytics mostly in the identification of microsegments? It's the sociology of program change that matters most. Later in this chapter, we provide an example about how identification data (segmentation) is combined with a program intervention to effect an important health outcome.

One could envision all types of predictive algorithms for many other population segments that would be relevant for population health management. What are the questions that were never asked because either the absence of data expunged the very thought or the customary mental models of an industry narrowed the scope of possibilities? For example, in the banking chapter, we think about a banking adaptation for healthcare that uses easily available personal data to make predictions about the challenges and capabilities of people to respond and carry through with treatment. Not in the biological sense, but in the social sense. We all have a FICO score, which is absolutely critical in getting a loan. The score is important to the lender because it indicates the person's capability to carry through with a promise made to repay a loan. In healthcare, it's not so much what the doctor does, it's what we do as people to adopt healthy behaviors. And sometimes all our wishes to behave in a healthy way are subverted because of other social factors that create barriers to do the right thing. But there is no measurement of a patient's willingness, capability, and social constraints to improve one's own health. And the measurements are easy...if one uses the right data, as is done in banking.

Lifetime Value to Concentrate Marketing Spend

Customer lifetime value (CLV) is a measure of the revenues/profits attributable to a given customer over her career with a company. It represents how much each customer is worth in monetary terms and how much a marketing department should spend to acquire and keep her based on the return on investment. In Table 3.1, the lifetime revenue and profit for hypothetical average customers versus the best customers is calculated.

Table 3.1 Lifetime Revenue for Average versus Best Customers

	Best Customers	Average Customers
Life expectancy	3 years	2 years
Revenue year 1	$280	$100
Revenue year 2	$280	$100
Revenue year 3	$280	$0
Lifetime revenue	**$840**	**$200**
Gross profit margin	10%	10%
Lifetime gross profit	$84	$20
Acquisition cost	$8	$4
Lifetime net profit	**$76**	**$16**

The bottom line is that the CLV for the best customers is almost five times greater than that for the average customer. Marketing, concerned about its ROI for its investments, would quite logically focus its investments on the best customers. For example, one retail customer analytics firm says, "We'll let you know how much a Groupon customer is worth and how customer value varies across ad networks and channels. And we let you know early, so you can take actions on the results."[24]

What are the implications for healthcare? CLV can be approached quite differently depending on a few parameters: actors (providers or payers); payment system (fee for service or global payments); value metric (dollars or clinical outcomes). For providers in the fee-for-service system, the prediction of medical utilization and revenues is relatively easy. The 80/20 rule applies; that is, 80% of utilization is attributable to 20% of patients. A quick calculation might suggest that retirees with multiple chronic diseases would produce the highest revenues under the fee-for-service system. However, Medicare reimbursement rates are lower than those of commercial insurance for the same services, and more volume might not produce better net profit. Similarly, certain procedures produce better profits because of payment formulas. For example, cardiac surgeries produce a high profit, especially when performed in "focused factories." And, of course, certain medical specialties produce better lifetime revenues than others.

But providing more services does not necessarily produce better health and there is growing concern about overtreatment. And as global payment approaches achieve more coverage and as transparency and risk adjustment methods level the playing field, lifetime value based on profitable volume will become less useful.

For payers, profits increase when the utilization/cost experience of members, factored into premiums, is less than predicted. Utilization can be moderated by good medical care, a focus on prevention and medication compliance, and disease management of chronic conditions. It can also be hedged by increasing the pool of healthy individuals to begin with.

The payer business model will change dramatically over the next few years as payers go retail through health insurance exchanges, perhaps doing most of their business directly with consumers rather than employers. Additionally, insurers will have to take all comers with no exclusions for preexisting conditions. Community rating, rather than experience rating, will be the norm.

This presents a marketing conundrum for health plans. Using CLV precepts from retail, insurers should identify the best, most profitable members. This segment is the young, healthy, and wise. After targeted marketing to enroll this group, the next strategy would be to keep them healthy through wellness programs and keep them loyal through good customer experiences.

Insurers can also disenfranchise the sick/costly members. Customer relationship marketing systems can provide information to deliver poor customer service and frustrate them in other ways, such as not paying on time, excessive Explanation of Benefits (EOB) notices, and harsh communication tactics. One marketing firm refers to disenfranchising strategies as "rooting out unproductive relationships."[25]

Skimming, or selecting the healthy, is easier, less expensive, more assured, and more immediate than medical management. Lifetimes are uncertain in healthcare. Some plans and providers believe that improving health outcomes through prevention and disease management efforts might not pay off for them although it might reduce costs for Medicare. Employers change insurance plans often. Also, there is low member loyalty to health plans, employees change/lose jobs, patients have low compliance rates with disease management, successful implementation can be difficult with independent doctors, and it takes a long time and costs a lot. Although skimming is the wise thing to do for quarterly earnings, it misses the mark on the primary obligation of healthcare: to save lives and improve health.

What if lifetime value were actually about the length of a life? What if health plans/systems were rewarded for producing longer and more functional lives? And what if there were financial incentives for this lifetime value rather than the production of more services? So the concept of lifetime value all turns on the value orientation. Using analytical methods for tracking people longitudinally over a lifetime would put the focus on long-term improvements in healthcare. Also, CLV analytics *can* be used to improve health for everyone, to lift all boats, and by doing so can add value to the bottom line and fortify the brand through its integrity and dedication of purpose to the health and well-being of all customers.

Market Matching: Who and How

Segmentation is about isolating distinct clusters of homogenous customers that can be characterized by certain buying patterns, for example, "brand loyalists," "social media-ites," and "bargain with me."

These clusters can be activated when trigger events occur, including moments that matter, inbound call content, changes in insurance coverage, spikes in medical costs, social media sentiment, and insights from advanced predictive analytics.

Marketing messages are crafted to fit the cluster's habit/buying receptor site in terms of language, argument, and specific offering; for example, coupons. The messages are delivered through the preferred channel, such as mobile device, outbound call, customer portal, e-mail, or face to face.

The major dependent variable (outcome) in the equation is CLV or specific revenue from targeted campaigns. The major independent variables are the cluster, channel, trigger, message type, and intervention. The precision of the equation can be improved by gathering more and more data on the individual to drive more segment differentiation.

How does this model apply to healthcare? First, the dependent variable needs to be specified. The model can be applied to attract new customers to sign up for a health plan or provider system (acquisition), to keep them in the fold as a loyal and satisfied customer (retention), to grow revenues through up-sell (more expensive health benefit plan, more medical services if FFS), or to cross-sell (from medical to disability insurance).

But the real benefit might come from changing customer behavior to improve health and functioning, especially in global payment settings. The goal is to encourage and reinforce customers as co-producers of health. The clusters become very different from retail because they need to predict health improvements. And the data needed to define them is very different than that in retail.

CVS/Caremark has multiple stakes in healthcare. It is a retailer of prescription medication and health products through its chain of CVS drug stores, a healthcare provider through its MinuteClinics, and a pharmacy managed care program through its Caremark pharmacy benefit program (PBM). It promotes a better healthcare system through its emphasis on medication adherence, which leads to improved health outcomes and lower costs. It improves its bottom line through a paradox of increased retail sales and decreased prescription costs to employers through its PBM.

Its Consumer Engagement Engine pools data from all customer touch points, including retail and specialty pharmacies, mail order, MinuteClinics, Web sites, the telephone pharmacy service, and the ExtraCare loyalty program. It develops a full picture of a person's prescription drug purchase and usage behavior, as well as related personal and healthcare information including demographics, medication history, prescribing physicians, drug usage, and so on. It identifies customers with suboptimal pharmacy care—for example, not taking medication—and those at risk for it. It then uses its Pharmacy Advisor program to change behavior regarding compliance by communicating with patients at the right time and with the right message. The right time is usually when patients need to renew medications and they come to a retail store or fulfill through mail order. The pharmacist has the information available at the point of service to counsel the patient on the importance of the medication and to answer any questions the patient might have.[26]

The results of the program, for patients with diabetes, show modest improvements in medication compliance of 2.1%. The rate was better with face-to-face contact with pharmacists through store consultations than with phone calls made by pharmacists through mail order.[27] They estimate that the total cost recovery from perfect medication compliance for all medical conditions in the U.S. would be $300 billion. So if their improvement percent were brought to scale, there would be savings of about $6 billion.

This example demonstrates a sophisticated use of analytics to collect and integrate customer data from all corners of the enterprise and apply the analytics for segmentation, multichannel matching, messaging, and interventions. It focuses on the laudable goals of saving lives and dollars through pharmacist interventions.

There are some limitations. The "over the counter" brief dialogue with a pharmacist, who is multitasking in doing her day job of filling prescriptions and other new jobs such as giving flu shots, might not be concentrated enough. And customer communications from PBMs have often been focused on switching to a cheaper drug and appearing to benefit the company's bottom line rather than the customer's health needs.

Also, pharmacists do not know the patient's medical history and might not have the knowledge or context to answer important medical questions. The pharmacist cannot substitute for the medical team. In fact, a big part of the problem with medication compliance is the high prevalence of side effects or no effects, and the difficulties in achieving a satisfactory level of communications with the medical team to address issues in a timely way before the patient becomes uncomfortable and stops taking the medication.[28] Of course, fragmentation of healthcare delivery systems and the high cost of some medications also inhibit compliance.

The obvious point is that if CVS can do this from its important, but limited, pharmacy perspective, why can't medical teams do it from an overall medical management perspective?

What if patients/members/customers were known more fully through more complete data than claims, enrollment, and medical records, and new clinical segments were defined beyond diagnosis, stage of disease, and other medical classifications to include propensity to engage, self-monitor, partner on shared decision making, and change behavior patterns? What if there were a radical change in the dependent variable to include patient/customer control and how to support it? What if this information on mass, radical customization to a person-of-one facilitated a change in the discourse to improve coordination of care and mining of EHRs, and to foster a true transformation in the role of the customer/patient?

Mobile and Social Media: Creating Buzz and Advocacy

Customers have a lot more information about attitudes, preferences, and suggestions in their heads than is evident from their data streams.

Mobile devices are the new nexus of communications among individuals and with businesses for commerce. As mentioned earlier, customers expect 24/7 omnichannel reliability. And customers have a new power through transparency for comparison shopping.

There is valuable data to be harvested through mobile devices, including clicks, sentiment, purchasing behavior, and more. But the curious thing is that most of the data from most industries including retail does not come directly from the customer. In fact, most is covert and is used without the customer's awareness or consent.

Mobile devices represent the ultimate in communication capabilities and most users love them; when polled, one-third said they would rather give up sex than give up their mobile phones.[29] The point is that using the device to simply scrape usage data might be useful but misses a larger opportunity. The reality is that customers have a lot more information about attitudes, preferences, and suggestions in their heads than is evident from their data streams.

Marketers realize that more data is better, but the usual ways of engaging with customers do not work well. For example, Internet retailers often ask for feedback on a product through a brief survey or requests for customers to post reviews, but not many customers feel the allegiance to a company/product to do this in a constructive way. Some surveyors are now paying for responses to surveys to induce better response rates. E-Rewards pays a modest amount, for example, $3.50 in loyalty points, to answer many questions on a wide variety of issues, from travel to electronics preferences.[30] Appreciating the customer's time by paying for it is a step in the right direction, but the process might feel like a survey mill that does not provide much connection between the company and the respondent. And some companies offer the "privilege" to be on an "insight panel," which feels like being on an advisory panel, but seems to be more their taking information than an exchange of mutual value.

New ways to get customer input are emerging. These can include extending a positive experience to customers by making the focus of communications fun, engaging, and dedicated to purpose. Instead of focusing on the few who will not wean themselves off brand-name drugs to save money, it can focus on the many who do go to generics and provide positive reinforcements to improve health; for example, with games that teach about health, and free smartphone apps that monitor health. These can be offered and won through a lottery or quiz. Health Prize is an example of such a company with a tag line of "Driving measurable medication adherence with rewards, education, and fun!"[31]

And traditional marketing has its limits. Marketers acknowledge that many customers do not trust their hard-sell tactics and are more likely to take the advice of friends and fellow shoppers. Shoppers increasingly depend on customer product reviews, for example, from TripAdvisor or Amazon, and feel a responsibility to reciprocate with peers with their own published reviews. New marketing strategies focus on inducing a "buzz" and accelerating online word of mouth through social media such as Twitter and Pinterest.[32] But this is not relinquishing marketing to the wisdom of the crowd entirely. Rather, it is smartly inducing the conversation and facilitating its spread.

How does this apply to healthcare? The traditional ways of collecting information from patients and members are not working well enough toward the critical goal of dramatically improving communication and coordination among providers, insurers, and customers. There is a reliance on what data are "under the lamppost," such as claims data, and not on what data are available outside the industry that might be of use, as in other industries. Moreover, there is a huge cache of data available directly from people/patients that goes untapped. This includes data on health risks, ongoing functioning and emerging symptoms, behavior patterns related to self-care, feedback on care received, and much more.

Patients are being asked about their experience of care in hospitals; however, these surveys reach a small minority of patients. But people are tired of filling out forms. They do this multiple times during the same encounter at a physician's office. They are reluctant to provide information to health plans because the trust level is low. People are tired of giving lots of data without being clear how about how it is being used.

Increasingly, people are circumventing the healthcare system in giving their feedback and are sharing it with friends on Internet sites, as they do when rating travel and merchandise. The challenge for healthcare is how to gather more data about members and patients, in order to service them better and help make them well, in a way that is mutually beneficial.

The answer might be complicated. What if the rules of the relationship were changed such that patients and families had much more say and the healthcare providers fully believed and practiced a cardinal rule about the use of data in healthcare: "No decisions about me are made without me"? What if the data collected were clearly linked to the patient's well-being and outcome rather than helping organizations improve their own processes and profits? What if care were truly patient-centric? What if care was less bureaucratic and people felt that providers had the time to address their concerns? What if insurers led by putting the member first and started by changing the primary mode of communications to members, the Explanation of Benefits letter, so that it was understandable and helpful? What if people were really invited in to co-produce health? It really is about renewing the trust in the relationship bolstered by demonstrable attention to the customers' needs that will increase information flow. And the key to communication about all these things might be the mobile phone.

Conclusions

Retailers are masters of customer analytics for marketing. The extreme challenges faced by the industry from the Great Recession, hypercompetition, and disruptive innovation from Internet sellers and digital technologies have resulted in increased attention to the customer. More than any other industry, retailers have developed leading-edge analytic techniques to capture the minds and wallets of customers. They have used transaction data from within the industry, clicks data from the Internet, purchased data on a wide range of personal attributes, and video recognition technologies to make tailored offers in real time.

Although retail is the highest-rated industry for customer experience, and healthcare is the lowest, the industry does not achieve this through close personal relationships with customers. Most marketing and much of sales are accomplished through mass customization. Retailers have refined this to the point that a store like Target can predict what you want to buy before you realize it and then provide the messages and incentives to close the deal.

There are four analytic sweet spots that have relevance to healthcare, including predictive modeling and the use of nontraditional data, customer lifetime value, market matching, and mobile phone devices, along with social media. Insights on the use of these methods in healthcare include using nonconventional data to predict obesity and other microsegments not usually ascertainable with within-industry data such as claims; using CLV, not to skim for the healthy and disenfranchise the sick, but to concentrate on producing longer and more functional lives by measuring lifetime functioning and paying for it accordingly; using data to understand behavior change opportunities during "moments that matter" for many health microsegments; and realizing that mobile is the platform for better communications, the collection of data that matter, and truly listening to customers.

4

Politics (Presidential Election Campaigns)

After Eisenhower, you couldn't win an election without radio. After Kennedy, you couldn't win an election without TV. Obama showed us the same thing for social media. But in 2012, it's the data that will make the difference.

—Strata Conference, Get the (Data) Vote Out, June 20, 2012

Introduction

Political campaigns have a very concise performance metric: *Win* the election at all costs. And the costs were high in 2012 for the presidential campaign. It is estimated that more than $2 billion was spent by the campaigns directly,[1] with hundreds of millions more spent by super PACs and other committees.[2] Much of the effort was targeted on the million or so swing voters from eight "tossup" states at a cost of well over $1,000 per vote. If $1,000 is the acquisition cost, what is the value of a vote? This acquisition cost is certainly a high mark for the industry and for other industries as well. Is a vote of value equivalent to a TV, car, boat, or house? Industries certainly compute what the return on investment (ROI) is for marketing and sales based on the value of the product. The value of an election is not clear... and the sky seems to be the limit.

The 2012 campaign might have been a turning point in how strategy, decisions, and operations are executed. The influence of larger-than-life political strategists/gurus, who direct the course of the campaigns based on their deep political intuition, might be ceding a bit or a lot to data-driven approaches. The old ways of image making, negative ads, and top-down campaign management are just not sufficient to win. The new way of knowing "each" voter through personal analytics is a surer route to victory. And in order to act on this newfound data, a new management approach was used that democratized the data to the individual campaign worker and relied on them to deploy it with their own energy and judgments to achieve their own dashboard metrics.

The prevailing opinion is that President Obama won the election by collecting and deploying big data to understand and track the opinions and voting inclinations of swing voters and then embed that information in campaign operations to get more votes. The Obama data operation, named "Narwhal" (for the arctic whale with a long spiral tusk), built an unprecedented data warehouse including data from pollsters, fundraisers, field workers, consumer databases, social media, mobile contacts, and voter files. The data was used to increase the effectiveness of campaign workers on the ground, social media outreach, and all other channels. The data-driven approach was so strong that David Axelrod, President Obama's political advisor, said that the biggest single lesson from the campaign to be remembered for 2016 was "I would invest in people...who understand where the technology is going and what the potential will be by 2016 for communications, for targeting, for mining data, to make precision possible in terms of both persuasion and mobilization."[3]

In addition to the stunning success of analytics to essentially guide who the leader of the free world would be, analytics was in the spotlight for its predictive accuracy on who would win the election. One analyst, Nate Silver, predicted the winner within a few electoral votes six months before the election, and this did not change much right up to the eve of the election despite all the campaign antics that ensued. He has also become a crossover star, going from being a cult hero "stathead" to a highly regarded commentator on politics and other industries and a best-selling author for his book on predictions, *The Signal and the Noise*.[4]

The 2012 presidential election is a glorious testament to the success of harnessing big data, predictive modeling, and an old standby, polling, to achieve a momentous outcome. There is just no comparison from other industries that is anywhere close to this stunning accomplishment of analytics.

In this chapter on analytics for the political campaign industry, we cover the following areas: What is the context of the industry in terms of its major strategic and operational challenges? What are the industry's sweet spots? What are the adaptations for healthcare?

Industry Challenges

Winning a presidential election in the U.S. is a huge challenge because the electorate is deeply partisan and was equally divided on its choice in 2012. More than 97% of the electorate had made up their minds. Only 3% percent were uncommitted. And only one or two percentage points would sway the election.

The fundamental strategies for getting votes cast in your favor on Election Day are these:

- Get enough swing voters to commit to your side...and make sure they vote.
- Hold the base...and make sure they vote.

Swing Votes

It's about the electoral count, not the popular vote. So work has to be distributed to every state, more or less, depending on the number of electoral votes at stake and the return on investment to win a given state. The swing votes matter in just a handful of states to shift the popular vote in one direction and put that state in the electoral column. Winning the election came down to just eight states that represented 95 electoral votes.

But not all swing votes are of equal importance and this makes the analytics interesting. It was possible to classify the probability for the swing voters in a given state to vote for Obama or Romney based on their personal characteristics and how "elastic" the state is in terms of how overall issues like new economic data influence the entire swing population. Undecided voters are not necessarily swing voters and many undecided citizens are not going to vote or swing. For example, African-American males are more likely to vote Dem. Female Southern Evangelicals are more likely to vote Rep.[5] Knowing more and more information about the individual and the correlation of that information with voting tendencies produces an increasingly refined prediction model of which way a swing voter will decide and whether they will go to the polls and vote. An exquisite logistics regression model developed by Nate Silver,[6] which incorporates an understanding of swing voters' characteristics and the overall elasticity in a given state, was able to provide a valid estimate of the likelihood that an individual voter would be influential in determining the Electoral College winner. The model produces a return on investment metric, which is the likelihood that an individual voter would determine the Electoral College winner. The highest ROI states were New Hampshire, Ohio, Nevada, and Virginia. Obama won all of them.

Swing voters, also known as fence sitters, can be easier to find than to persuade. They do not watch Fox or MSNBC and are generally AWOL from most things related to the election. For example, for those without a college degree, those who are trying to get one, and those who are working two jobs to stay afloat in hard economic times, the presidential election is just not on their radar screen. According to Jeffrey Pollock, the president of Global Strategy Group (a polling firm for Priorities USA Action, a pro-Obama super PAC), white noncollege swing voters are "particularly low-information voters who don't pay attention to the daily political back-and-forth, so their opinions are driven by their economic situation."[7]

Fence sitters, by definition, do not show commitments easily, so it is hard to know how to influence them. Some say it is about emotional intangibles that might trigger their aligning with one candidate or the other. Words that cause a trigger include female healthcare, Medicare, abortion, jobs, "government run," socialism, the 47% crowd, lower taxes for the wealthy, and so on. And these issues got a lot of play in the 2012 election. For the fence sitters, it can well come down to a single event like a debate, the jobless number in November, or an unanticipated world event, like an eruption in Afghanistan. Perhaps the emotional trigger issue and the paucity of time to be fully engaged are indeed the reason why negative ads were so much a part of the campaign. In addition to the debates and extensive advertising, there will be boots on the ground to hunt, find, convince, and knock these important voters off the fence and into their cars to drive them to the polls.

Hold the Base

The other, complementary strategy is to get out the vote from the core loyalists. Conventional wisdom on elections strategy says that when there are few persuadable voters in the middle, it comes down to a turn-out battle. This strategy requires constant pressure in each state, especially in those that are "leaning" one way or the other. There are ten states leaning. If you live in died-in-the-wool red or blue states—for example, all West Coast states are for Dems and all Southeast states are for Reps—you might never have seen the endless cascade of political ads in the 2012 election. The electoral vote for these states is considered a done deal. If you live in Ohio or Florida, the states with the highest electoral votes in play, you might have very well seen the President up close and personal at your neighbor's house.

Industry Strengths

To understand how analytics contributes to the previous strategies, the strategies need to be decomposed into the key tactics that support them. There are seven key strategies, all starting with *m*, including money, messaging, medium, microsegmentation, math, mobilization, and motivation.

Money

Money raised from supporters is the essential fuel for political campaigns. Without it, the campaign comes to a halt (as it did for many candidates in the Republican primaries). And not enough of it can tilt the scales through an imbalance of TV ads for the more wealthy side. The amount of money raised for this presidential campaign exceeded $3 billion.

Most of the funds raised for the 2012 presidential campaign did not come from a broad representation of citizens. Contributions to the campaign coffers of less than $200 amounted to about 40% of the Obama campaign and 20% of the Romney campaign (as of the end of August).[8] However, most of the money spent on the presidential campaigns will come from super PACs that were spawned by the U.S. Supreme Court's Citizens United decision. The super PACs cannot provide funds to the campaigns and must conduct their spending independently. There are no legal limits on the amount of money that can be raised or from whom, including corporations, unions, and foreign governments. This can severely limit "full representation" in a democracy and the real possibility of buying the presidency.

Most industries raise capital from financial markets or use earnings to fund operations, new product development, and all the rest. Political campaigns are different and are more like charitable organizations than profit-making organizations. The latter does not need widespread public approval (unless it affects the brand and sales). It mostly focuses on shareholder approval, which primarily rests on good financial results. The former depends on the generosity of followers and powerful influencers to exist at all. In some cases, politician millionaires like Mayor Bloomberg of New York City and Governor Rick Scott of Florida provide a major source of financing. Usually, the campaign must raise mostly all of its funding through solicitation.

There are a few unique features about raising funds in the political sector:

- Of most importance is that people need to *believe in the cause*. The cause must tug at the heartstrings. It must evoke emotion, the preservation of an ideal, a change in the status quo. It is doubtful that a singular approach, like incessant daily e-mails, however well executed, will work. (And the Obama election campaign got annoyance demerits for its daily onslaught of e-mails from the President, VP, First Lady, and others asking for money.)
- The campaign must continually engage. The events of the world, and negative TV ads, have dramatic effects on voter sentiment and perceptions of their place in the world and their view of the candidate. Therefore, it is necessary to keep followers in the fold.
- Politics must touch a lot of people to raise funds, including the small and big supporters and all those in between. As with other businesses, it has different price points for different buyers.
- Finally, there are different legal and ethical constraints related to raising money in politics relative to business. A privacy breach about donors in politics would engender immediate press coverage. There are constraints on the maximum direct contributions to campaigns. There is much more

transparency and media activity to expose ethical and legal lapses. Eyes are always on the campaign and there is no place to hide; consider, for example, campaign troops attending speeches with their minicams looking for a newsworthy gaff.

These features define how analytics supports one of the most important tactics of politics, raising money. On the one hand, raising money in politics is similar to making money in corporations, for example, understanding markets and various buying segments, targeting messages, providing purchasing incentives, and delivering on the promise to earn repeat business. On the other hand, it is quite different in that the representation of the product needs to be constantly updated to respond to opponent attacks and events largely out of its control, as well as the fickleness of an electorate that can change its mind because of a grimace or gaff in a debate. Hence the need for the next *m*, effective and continually evolving messaging.

Messaging

Messaging is a refined art and science in politics. For example, Vice President Biden's pithy response to the Republicans' question, "Are you better off today than four years ago," was this: "Bin Laden is dead and General Motors is alive." It was concise and held a lot of meaning that responded to deep concerns of its intended audience. Also, politicians are on the stump every day, prone to many questions (and videos) on their positions, and "staying on message" is vital.

There are five essentials for an effective political message: (1) frame the issue, (2) appeal to values, (3) keep it simple, (4) keep it believable, and (5) create a mental picture.

Messages need to be different for different voter segments. For example, a tax message to one group might be, "We need to invest in infrastructure, education, and so on." Alternatively, the tax message could be, "Americans need tax relief." But the reality is that there are many segments, and the more targeted the message, the better. It is conceivable that there could be as many message packets as there are microsegments. In fact, it could be tailored to a market of one. Analytics contributes fundamentally by defining distinct microsegments based on data about values, what images resonate, what language works, and so forth. And this requires a good deal more data than is currently available.

Medium

Many media channels are available to get the message out, including newspapers and magazines, flyers, billboards, e-mails, Web sites, TV, voice phone, smartphone, social media, and more. People have their own preferences on how to receive information. It is important to match these channel preferences to targeted messages for specific microsegments. In general and across industries, wholesale approaches such as 30-minute TV ads and voice-phone messaging are giving way to customized approaches such as changing ads on a mobile device, which takes into account voter file data and the "cookies" placed on consumers' computers by commercial advertisers. Increasingly, mobile has become the nexus of the multichannel portfolio.

Microsegmentation

One of the key lessons from the Obama victory in 2008 was "divide and conquer."[9] That is, divide up the general population into smaller segments and do unique and compelling things to change their behavior. Microsegmentation is a central theme in marketing in all industries. Retailers have segments based on buying patterns like "brand fanatics," "let's bargain," and "media socialites." Healthcare has segments like "empty nesters" and "young and healthy." Politicians have segments based on voter demographics like "white, Southern, male, non–college educated" and "female, Evangelical, Southern." The challenge for microsegmentation is approaching the ideal of a market of one without the one-on-one friction and cost, and achieving it through more efficient mass, radical customization. More data on the individual leads to more fine-tuned methods.

Math

Data-driven enterprises require big data. And big data requires big math. The finer data granularity needed for modeling and advanced decision support requires much more data, better equations, and out-of-the-box modeling.

Polling has been the mainstay and essential analytical feature of campaigns for decades. Math is needed to make generalizations from samples and always includes the ever-present "margin of error." But polling, in and of itself, is yesterday's news and provides a static snapshot in time. More dynamic views are needed and that is where big math steps in, especially focusing on predicting the future. Modeling is an important math discipline that often results in predictions such as the likelihood of a fence sitter making a difference in electoral votes as discussed earlier.

Politics has captured many streams of new data, ranging from "listening" files from campaign workers, to social media, to personal data bought from data aggregators like Acxiom and Experian. Big math is needed to make sense of it, running all types of correlation tests to see how it can be used for predictions. For example, how much weight does one put on religious preference, home value, occupation, recent retail purchases, preferred brands, ZIP Code, donation history, and facial characteristics? The list of possibilities is endless.

Mobilization

Mobilization has to do primarily with getting voters convinced, motivated, and delivered to the polling place. Mobilization is the work of all the campaign workers on the ground in campaign offices throughout the nation, doing such things as knocking on doors, distributing leaflets, and the like. Mobilization is increasingly being accomplished over the Web and through social media sites to create and sustain followers and advocates.

Motivation

Presidential campaigns are long, two-year sprints. This is about motivating citizens to get engaged and vote. It is also about getting campaign workers to stay motivated and work to peak performance. Supporters leverage the key purposes and resources of the campaign at the ground level. Motivation goes beyond blunt, anonymous campaign messages and increasingly has more to do with relationship management. With the new generation of data-driven campaign workers, it is necessary to motivate them, including giving power (and all the information) to them and giving them the autonomy and attendant satisfaction to make their own decisions and take risks and develop networks of influence.

Analytics Sweet Spots

Politics incorporates a wide span of analytics, including most of the analytics across industries described in the introduction. Its unique analytic sweet spots that have relevance to health are polling, predictions (voting and winning), and new data (democratization) (see Figure 4.1).

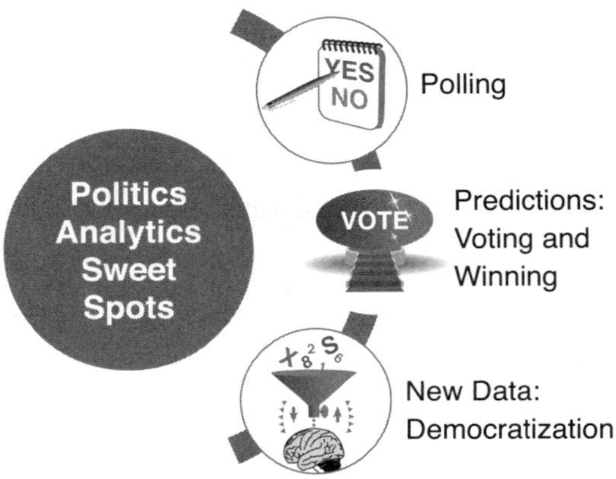

Figure 4.1 Politics analytics sweet spots.

Polling

This is a case of getting the right data directly from a willing participant's head rather than getting convenient data of questionable utility covertly from her data streams.

Polling is the quintessential mainstay of political campaign analytics. Polls seek to understand who is winning and why, and polling uses the scientific method to pick representative samples of voters, conduct interviews, and use statistics to achieve confidence in its results. It even makes the actual results of elections an afterthought because elections are "declared" early from exit polls, and even from polls conducted months in advance. In fact, polling was so accurate in the 2012 election that the actual act of going to the polls to vote could be considered a quaint ritual in the years ahead. Finally, polls captivate the electorate on who's winning the race. And the data provide critical feedback to the campaigns to readjust tactics.

Polling is a form of surveying that is done in every industry. Most notably, industries use surveys to get customer feedback on the satisfaction with products and services. Surveys are an important method for scientific research in health and other industries like life sciences and education. But political surveying has unique characteristics.

Polling is widely used, especially during campaign seasons. For example, five different polls were released on one day, for one state (Ohio), on the eve of the presidential election. Polling is highly accurate, especially when multiple polls are combined. The margin of error on the five Ohio polls was +/–1.5%. Polls are extraordinarily visible, appearing in multiple news outlets, often as the lead story. And polls are repeated over time for the same population (although not the same sample) and provide a longitudinal view of the data.

Another unique feature of polling and surveying as an analytic method is the way the data are collected, that is, directly from citizens, customers, and patients, as compared to data extracted from transactions such as purchases or medical bills or social media communications. This is a case of getting the right data directly from a willing participant's head rather than getting convenient data of questionable utility covertly from her data streams. The data is richer and more pertinent to the hypothesis under consideration; for example, who will win the election. And it has the potential to enable an ongoing conversation with the individual customer or citizen to gather even more data and to be customer-centric in the process—provided that the organization doing the surveying is trusted and gives something of value in return.

There are challenges with political polling mostly having to do with representativeness of the samples. Response rates are low (at about 20%) and declining. Telephone land lines are decreasing, and it is more difficult to conduct an interview with cellphones because laws prohibit "robocalling" and calling during specific hours. Additionally, cellphones can have an area code that is not where the respondent lives, thus confusing geographic sampling. Online polls are entering the mix of polling media. But, again, those who respond are Internet savvy and not representative either. Finally, social media sentiment analysis that quantifies the level and change in positive and negative messages from Twitter and Facebook are emerging, but the segments that "tweet" are not representative. The bottom line is that no one channel is representative; however, this is correctable to some extent and for some purposes through the use of predictive modeling, as described later in this chapter. And more data, from more sources, nearly always produces better accuracy through higher sample sizes and convergence.

About 60% of eligible voters went to the polls in 2008 and 2012. Their voices were heard and they were a part of something larger than themselves in determining the leadership and fate of the nation. A lot was at stake and the majority of Americans wanted to participate.

Does the same thing happen in healthcare? Not by a long shot. Even though the economic fate of the nation is very much determined by decisions about healthcare's performance as a social system, people are only remotely engaged in giving their voice for change. And, even more remarkable, people are generally removed and unengaged from most aspects of ensuring their own health...never mind the health of the country.

How might polling in healthcare help? There are entrenched obstacles. On the one hand, it is not clear that anybody is really listening to patient input through traditional healthcare surveying. And on the other hand, it is not clear what our roles should be as patients.

Patients do give feedback on their satisfaction with hospital care. Hospitals have been conducting surveys of patients on their satisfaction with care for a long time. They have used survey instruments from well-established commercial firms such as Press Ganey, and many hospitals have used the results to make improvements in hospitality and communications functions. Now, CMS is emphasizing the importance of patient engagement through a pay-for-performance initiative with about $1 billion at stake that rewards/penalizes hospitals based on the results of HCAHPS surveys (Hospital Consumer Assessment of Healthcare Providers and Systems). HCAHPS is an important advance in measurement because of the content (satisfaction and engagement), the substantial research by the Agency for Healthcare Research and Quality (AHRQ) to refine the tool, the standardization and implementation protocol that allows meaningful comparisons among hospitals, and public reporting on the CMS Hospital Compare Web site and others.

This is definitely an important step in the right direction, although many providers balk at the idea of "making patients happy when they have stage 4 cancer."[10] What's important to know is that a very small percentage of patients are sampled.

Hospitals, on the average, must complete only 100 patient surveys a year. This comes down to about 1.5% of hospital discharges in the U.S. (There were 36 million discharges in the U.S. in 2009[11] from about 5,750 hospitals.[12]) This sampling frame is adequate for making comparisons among hospitals presuming that patients are sampled similarly across hospitals. But it is notable that the HCAHPS participation rate is about one-fiftieth the size of voter participation in national elections. If we lament the poor turnout in voting, how should we regard the extraordinarily poor opportunity for voice in healthcare? (Note also that the HCAHPS public reporting and pay for performance is only for hospitals and excludes the vast majority of Americans who do not go to the hospital.)

Clearly, the purpose of HCAHPS is not to engage the public about the future course of healthcare policy. The past two national elections have certainly been about healthcare reform to some extent, and people registered their vote about it. The one issue of healthcare, however, was enmeshed in many other issues, most notably jobs and the economy. And it is questionable how much the electorate was actually engaged and understood the big issues. By and large, the "conversation" was a media "push" of carefully crafted messages and advertising manipulation to tell people how to vote. Can there, should there, be a dialogue about healthcare, especially in regard to the trade-offs that citizens as stakeholders should make that counterbalances the influence of vested interest groups?

The lack of opportunity for participation in healthcare goes much deeper than (just) 100 patients per hospital given the opportunity to provide feedback on their hospital stay. It also goes to the core of how people are engaged in their own treatments. People are the most important determiners of their own health. The research is very clear that personal behavior is three to four times more important than healthcare in producing good health. But it is also clear that people do not "comply" with treatment or "adhere" to prescriptions. And patients are not engaged in decision making where their preferences are most important, especially when considering elective procedures that have equivocal medical outcomes. This general lack of acceptance and practice of shared, or informed, decision making is another signal of the need for transforming the ways to engage people. It's one thing not to vote and make your voice heard in Washington. It's quite another thing to be absent from decisions about your own life and death. Similarly, the excellent movement to have patient-centered medical homes is very important to improve the coordination of care, but it is not the answer to achieving a patient/people-centric health system with optimal patient engagement.

Finally, only a small minority of people/consumers use "standardized" information like HCAHPS to select hospitals or health plans for a variety of reasons, not the least of which is that they do not have a whole lot of choice. They depend on their doctors to make referrals to hospitals and on their employers to select insurance plans for them. People do not act like the rational consumers that markets espouse, because, er, healthcare is a poor market. And it is pretty widely accepted that public reporting of performance information is directed much more at providers than at people. Providers do respond to the data even though the rationale for doing it is for "consumer decision making."

People who have health issues *do* want to talk about them. In fact, for many older people, health is a major agenda in their lives. It can occupy much of their time with frequent and regular medical visits. And the uncertainty and worry about the life-and-death consequences are at the forefront of their attention with seemingly an unquenched appetite for discussion, understanding, and resolution. This yearning for answers and guidance is not always provided by the medical system, and people remain frustrated by the incomplete responses to their questions and concerns.

The medical system might simply not be capable of providing the level of communication and engagement that many people want. And because they have not gotten it, they have receded from behaving in ways to elicit it because it has just not reinforced their needs. It should be no surprise to anyone that the medical system is fragmented and uncoordinated and lacks good skills in communication. For example, how successful have you been in having records transferred from one healthcare provider to another? In getting lab results or biopsy results in a timely way? In just trying to have a doctor call you back? How would a consumer react if this transaction and communications failure happened routinely in banking—for example, if deposits went to the wrong account, if the bank could not post a deposit for a week, or if it did not know what the balance in the account was and would not provide cash at the ATM? It doesn't happen in banking and it shouldn't happen in healthcare.

Clearly, the medical system can and should make improvements in communications, but it should not be the only avenue for patient voice. More and more people are favoring connections and communications with their peers and depending less on pronouncements from systems with their seemingly inherent obstacles to communicating with customers. Good customer service is more of a surprise than an expectation in most industries today. People want to hear about what friends say about a hotel, not what the hotel marketers are saying. People want to evaluate products and services and "pass it forward" for others in the community to benefit from. The idea of top-down surveys might feel like a top-down extraction of information rather than engagement. It's time for a change to get people engaged in healthcare. Politics and polling provide some clues.

It is important to recall that there was a time in the politics of the U.S. when citizen participation was a mainstay in shaping social issues and solutions. This was during the gilded age of community planning and it was particularly so for health. One important example was the Oregon Health Plan in the 1970s and 1980s, which included a great deal of citizen deliberation about priorities for healthcare, including what services would be paid for under Medicaid. It was the right approach to engage citizens in making trade-offs. Indeed, the only way to control costs might be to understand that resources are limited and trade-offs need to be made and that the politic process needs to activate deep citizen participation. Community planning and citizen engagement peaked during the Great Society era enabled by the legislation driven by President Lyndon Johnson. But citizen participation was, by and large, driven off the cliff when a market-oriented approach to healthcare, as ushered in by President Reagan–era policies, surged ahead of a government/community approach. Perhaps the pendulum has swung back a bit.

Why not have a "moon shot" on one big outcome in healthcare, like a presidential election? Get everybody involved, measure success, make improvements, and rejoice in victories? That one big outcome could be the well-being of citizens in the country. European and Asian countries and Canada are very clear that the success of a country is based not solely on its GDP but on the well-being of its citizens as well. And these countries believe that government should respond to the signals from its citizen customers and be accountable for improving well-being. These signals are measured and reported through surveys on well-being.

The Canadian Index of Well-Being (CIW) is a single composite measure derived from eight measurement domains (see Figure 4.2).[13] These include measures related to "healthy populations," as well as other determinants of well-being, including education, living standards, community vitality, democratic engagement, time use, leisure/culture, and environment.

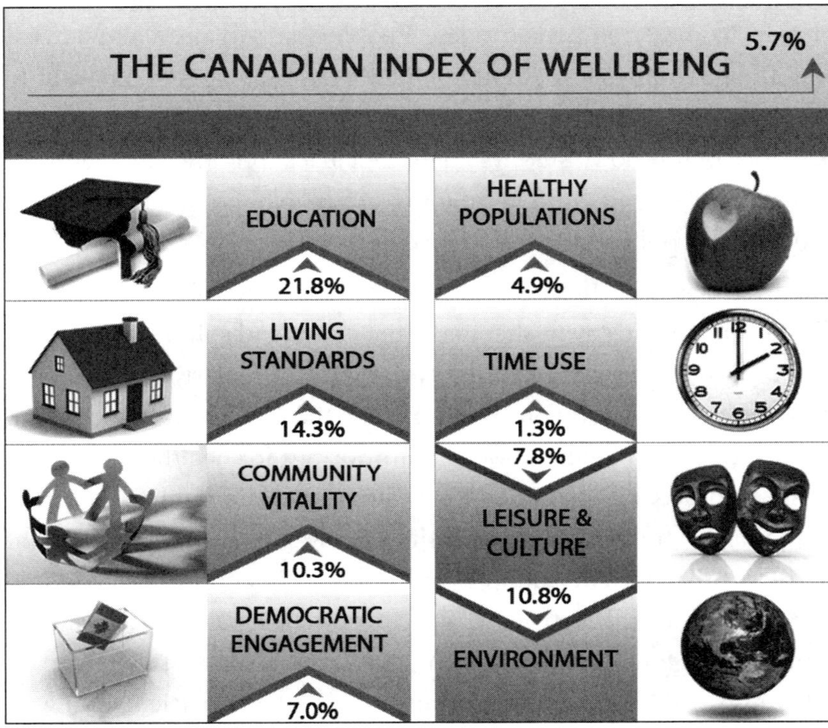

Source: Canadian Index of Well-Being (2012). Reprinted with permission from the Canadian Index of Well-Being.

Figure 4.2 Components and growth rates of the Canadian Index of Well-Being.

The percentages in the figure are the changes in the index from 1994 to 2010. So the 5.7% change in overall well-being represents a very modest annual change of less than 0.4%. During the same period, the economy, as measured by the GDP, increased 28.9%, or about 1.7% per year (see Figure 4.3). The improvements in GDP are not well correlated with improvements in CIW. For example, for the first ten years in the chart (until 2003), the CIW remained flat although GDP increased 23%.

Note also that China measures well-being, or happiness. The China index also shows the same lack of correlation between growth in GDP and growth in well-being.

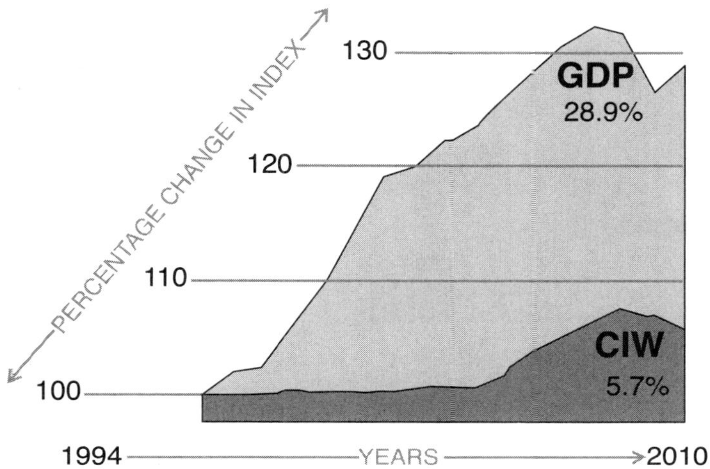

Source: Canadian Index of Well-Being (2012). Reprinted with permission from the Canadian Index of Well-Being.

Figure 4.3 Percentage change in GDP and CWI.

The "healthy populations" domain includes measures on smoking, immunization, life expectancy, probability of depression, and good health years remaining (see Figure 4.4). Some measures require citizen participation through Statistics Canada surveys to populate the metrics, including patient ratings of health services, self-reported diabetes, and self-rated overall health.

Note that the improvement rate over the 17 years is 4.9% and the major positive change included a drop in smoking among teens of 85%, which was partially offset by an increase in self-reported diabetes of 53.1%.

In the U.S., there is a good deal of surveying on health and healthcare. In terms of health, multifaceted measures of health at the county level, called County Health Rankings,[14] are publicly available. And comprehensive views of quality of healthcare based in part on population surveys are reported at the national and state levels by AHRQ and the Commonwealth Fund. But the evidence is not compelling that the information in these very good reports is being translated into action by governments and health systems responsible for the well-being of their populations. And, again, it is more aimed at the policy and provider professionals than to the individual citizens.

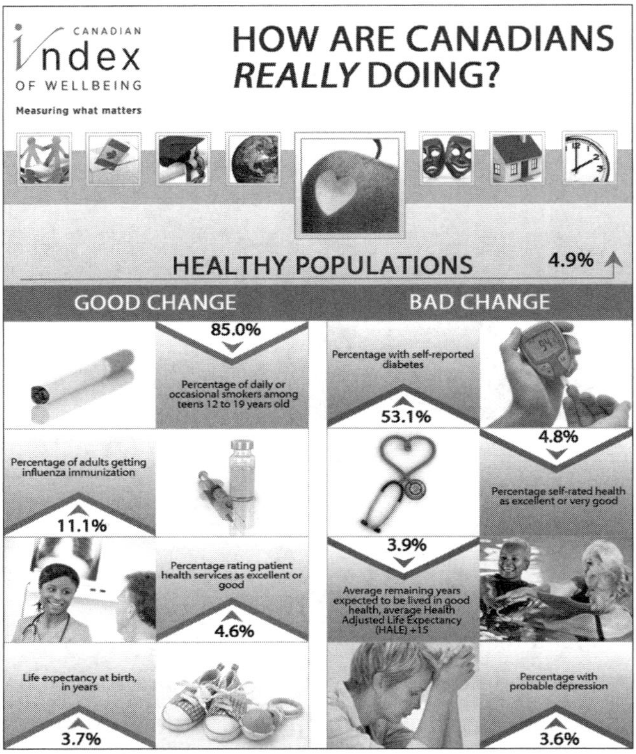

Source: Canadian Index of Well-Being (2012). Reprinted with permission from the Canadian Index of Well-Being.

Figure 4.4 CWI: healthy populations.

The learning from political polling might boil down to this: *Keep it simple, keep it focused, make it public, and make it meaningful.* In Canada, the provinces "compete" to be recognized as the best place for citizen well-being. Alberta is the place to live in Canada, according to the CIW. And they use it to improve government services to its citizens. Ideally, the index would be used to hold political figures accountable for the well-being of those they serve, as they are for education, roads, and jobs.

In the U.S., the technology exists to have a meaningful well-being index at the national, state, and local level. Underlying much of the surveying of people about their health in the U.S. is CDC's Behavioral Risk Factor Surveillance System, commonly referred to as BRFSS.[15] The CDC conducts 400,000 interviews a year. It includes 16 core sections that are asked of all respondents, including questions on health status, healthy days, and risk factors such as exercise, smoking, and blood pressure. It also has 34 optional sections, including sections on managing chronic diseases such as diabetes, heart attack, and depression. It also includes questions on related but nonmedical issues, including reactions to race, for example. BRFSS is usually reported at the state level. It can also "oversample" a geographic area such as a city for a more granular read on health. Cities and towns, rather than states, regions, and nations, are more localized and amenable to community engagement and action, and that is where the focus should be on measurement. The bottom line is that BRFSS is an amazing resource for "polling" people about health and well-being in geographic units that are most meaningful to them.

An example of oversampling at the city level is taken from the work of the Public Health Department for the Commonwealth of Massachusetts.[16] The state paid for the oversampling of large, relatively poorer cities on a few survey questions of the BRFSS. The results for one of these questions, which asked respondents to rate their overall health as excellent, good, fair, or poor, are included in Figure 4.5.

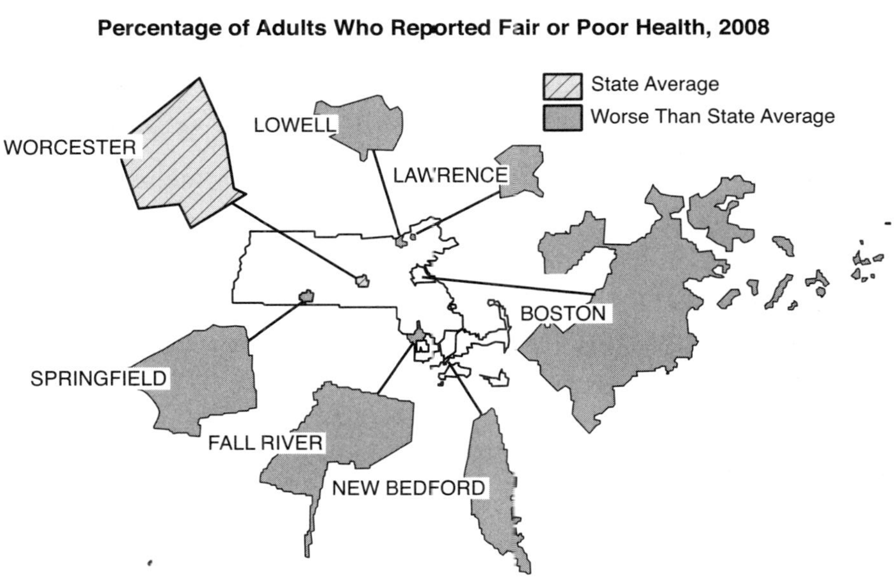

Source: The Massachusetts Department of Public Health, 2009. Reprinted with permission from the Massachusetts Department of Public Health.

Figure 4.5 Health ratings by cities in Massachusetts.

The chart shows the percentage of respondents who reported fair or poor health. One city, Lawrence, had 30% of its citizens reporting fair or poor health, compared with the state average of 15%. New Bedford, also a relatively poor city, had a rate of 26%. These are compelling data to show a marked (2:1) difference in perceived health that should warrant some action.

The problem with taking action might be that the technology overhangs the sociology (yet again!). It is possible to poll people about their perceptions of health and to achieve a good deal of scientific accuracy. (Note that some researchers perpetuate inaction by questioning the validity of self-reported health assessments.) However, the sociology might not be in place to do anything with the information, therefore rendering the measurement process relatively useless. In the State of Massachusetts, the government's role was to inform and educate through its publications and public presentations. Alternative action might include state aid or organizing community groups to improve health or public hearings to understand, explain, set an agenda, and coalesce community action for improvement. It might also include a public reporting and pay-for-performance program to reward municipalities that are making progress. In all fairness to the state, and to other states, the budget crisis over the past few years resulted in very significant reductions in funding for public health programs.

So the point here is that a survey needs to have meaning for the respondents. It should engage people beyond just producing answers. The topics need to be of interest to the people. It has to be simple. Preferably, it rallies a community's interest. Greater use of BRFSS and a moon race on well-being are good possibilities. But the process of engagement and the ways the data are produced might be anachronistic. More on this in the "New Data: Democratization" section.

Predictions: Voting and Winning

The data derived from political polls are an important element for predictive modeling about likely winners. But there is much more involved in making accurate predictions. It is instructive to look at Nate Silver's *FiveThirtyEight Methodology*[17] to understand how sophisticated it is. There are seven steps:

1. Determine the weighted polling average across all available polls, which takes into account the recency, sample size, and pollster accuracy rating of each poll.
2. Adjust the polling average based on trend lines, house effects, and likely voter status.
3. Perform regression analysis of nonpoll factors (when there are few polls) that incorporates information on partisan voting index, composition of party identification in the state's electorate, sum on individual contributions received, incumbency status, recent approval and favorability ratings, and stature.
4. Take a "snapshot" by including the regression results with the weighted polling average to arrive at a new polling average.
5. Make an initial projection that takes into account distributing the undecided vote, adjusting large leads due to likely regression to the mean, and adjustments that current standing might not be the most accurate forecast.
6. Make adjustments to the projections based on other factors related to possible errors including when polls disagree, there is a larger number of undecided voters, and how far the poll is from Election Day.
7. Do a series of simulations that include decomposing the error associated with a forecast into local and national components and then randomly generating the errors over the course of 100,000 simulation runs.

The FiveThirtyEight predictions were highly accurate. On the eve of the election, the method predicted that Obama had an 83.7% probability of winning with 305 electoral votes. The actual number of Electoral College votes for President Obama was 327. What is even more intriguing is that the prediction of electoral votes was just about the same as was generated five months earlier despite all the campaign activities that ensued.

As a brief aside, the accuracy and methodology of polling could make voting obsolete. For example, polling can ensure representation from many segments, not just by states for the Electoral College determination. The Electoral College is a requirement going back hundreds of years to the Founding Fathers. Having to go the polls is inefficient and exhausting. Some voters in Florida waited six hours. Some subgroups are less likely to get to the polls due to transportation, work, and health issues. In addition to ensuring better representation, polling could be no less meaningful for citizen participation than voting. Polls can be conducted in different ways, including in person, by mail, and via the Internet. And polling can enable an ongoing conversation and ask more questions, more efficiently. Why should citizens have input on national issues (only) once every other year?

Predictive modeling in politics is used for more than projecting a winner. It has been used extensively across industries and in academic research, but only recently in campaign politics. It started in the 2004 presidential campaign and took off in the 2008 campaign. Obama's analysts built predictive models to characterize different types of voters by their likelihood to vote for him and what their views were likely to be on issues such as pro-choice or pro-life. A vendor, TargetPoint,[18] looks at the aggregate effect of a chain of variables on the likelihood to vote Republican. For example, if you add male, 59 years of age, and white to the formula, that bumps up the probability. If you add that the voter drives an SUV, that bumps it up again. And if the person has a Bible in the household, that bumps it up to 85% likely to vote Republican.

Predictive modeling can also be used for campaign operations, such as raising money, by understanding, for example, what type of individual is likely to unsubscribe from e-mail as related to different types of solicitations. This can lead the campaign not to ask for money, but perhaps only to ask for information, from certain individuals and thereby not suffer a disconnect from the voter. When it comes to sensitive subjects like contraception, a campaign could rely on its extensive predictive models of individual attitudes and preferences to find friendly recipients. For example, it might mean pulling e-mail addresses only for those who had identified themselves as women on their registration forms and whose voter records included a flag marking them as likely pro–abortion rights.

Other models go from predicting the behavior of groups of individuals to targeting the individual. For example, one vendor, Constituent Cloud,[19] says that its proprietary algorithm ranks "each and every voter in your district" on "voters that will help you win" using data such as party, voting history, geography, race, and sex.[20] These data are visualized. One vendor, Moonshadow, promises to "visually 'fly over' your state, viewing in great detail the location of every voter in your selection or 'zoom down' to the rooftop level to get information about the occupants of a single household."[21] With the right data and the right math and models, it is possible to go to a state and have a list of citizens, addresses, and their profile for a canvas worker to start a targeted house-to-house conversion process.

Predictive modeling works better with more data, and more data about the individual enables better targeting and customization. Political campaigns are relatively deprived of an ongoing stream of data about their clients (voters). Retail has mountains of purchasing data, banking has money transactions data, healthcare has claims and electronic medical records data, and sports has ongoing player performance data. Politics does have traditional data sources from voter registration, exit polling, opinion polling, and donation history data, but the periodicity is just every few years. And, like most industries, it wants more information on the individual. In fact, organizations like NationBuilder[22] provide free access to voter files to anyone.

One way to get more data is through "microlistening." One shortcoming of polling is that it cannot pick up on the intensity or priorities of voters' positions. The Obama campaign used "microlistening" to dig deeper on these issues. The voter's words are often included in the canvaser's notes, call-center phone conversation, or "share your story" on the Web site. The campaign used text analytics to unlock the unstructured data so that it could be integrated with other structured data.[23]

The campaigns also used the information that came from campaign workers and interested supporters. For example, when people registered for Obama's Dashboard, it asked them for permission to add information. This included such data as their name, picture, gender, networks, list of friends, and "any other information" left open for sharing on Facebook. It also asked for information on birthday date and city. And it requested approval to send e-mail and post to each respondent's Facebook wall.[24]

Political campaigns are collecting information about individuals without their knowledge or permission as well. Some of it is provided "implicitly," for example, through their clicks on Web sites picked up by "cookies" that track every move. Campaigns are also tapping information available from data purveyors such as Acxiom,[25] Experian,[26] Equifax,[27] and Spokeo.[28] According to the *New York Times*, the campaigns spent over $13 million on acquisition of data like "whether voters may have visited pornography Web sites, have homes in foreclosure, are more prone to drink Michelob Ultra than Corona or have gay friends or enjoy expensive vacations."[29] Consultants to both campaigns confirmed that they had purchased data from companies on "voters' shopping histories, gambling tendencies, interest in get-rich-quick schemes, dating preferences and financial problems."[30] The political director of the Romney campaign mentioned how the retailer Target anticipates habits and "we're doing the same thing with how people vote."[31]

In summary, political campaigns gather lots of information about voters, directly from them through polling, "listening," and campaign-worker interviews, and indirectly about them through clicks and purchased data. Together with better and cheaper computing power and more expertise in the data sciences, including predictive modeling especially, the campaigns achieved a sea change in capabilities, which might have spelled the difference in winning the election. According to the Obama campaign's data director, they gained valuable insights about voters' "motivations, attitudes, protestations...allowing campaigns to finally see through the fog of the crowd and *engage voters one by one*."[32] Isn't that exactly what healthcare needs to do?

Healthcare needs to predict and connect better. Its predictions are usually about which treatments work best for a given diagnosis and what course the treatments will take, including life expectancy (prognosis). But new data outside the usual bounds of available healthcare data, such as claims and member enrollment, offer new opportunities in predictive modeling. For example, adaptations for healthcare from the banking industry include the Healthcare Improvement Capability Score (HICS) on how patients respond and contribute to prevention and treatment through their own behaviors, social assets, and attitudes. Predictive modeling in this way can engage patients one by one by knowing how they are likely to respond and what messages are best for them. For example, an older woman, like my mother, who has a specific pattern of visits to her doctor, an amalgam of conditions, certain behavior patterns within healthcare and outside it, certain social assets, and her own attitudes and motivations, would respond to specific messages and treatments and not to others. Healthcare can certainly predict how "she will vote" about her health.

Another example of using personal data outside the normal purview of healthcare, derived from politics, is convincing the "fence sitters" to become a part of the healthcare system. These are people who do not engage the healthcare system or get in too late, and suffer consequences of illnesses that might have been prevented or treated earlier. They need to be known and proactively courted, just as in politics. But finding them might not simply be a function of currently available data. Predictive modeling using a more comprehensive data set might discover the characteristics of this microsegment and lead to targeting campaigns to "bring them in."

Another example of using personal data is the prediction of the likelihood of obesity. Obesity is an overwhelming epidemic in the U.S. Medical treatment is difficult because it involves much more than illness care. It requires behavioral change and social supports outside the medical boundaries, including availability of nutritious foods, transportation, and more. It is preferable to get people into the healthcare system before illness onset by identifying the likelihood of obesity. The most important indicator of obesity is the body mass index (BMI). It is very simple to calculate and requires just two pieces of data: height and weight. Yet health plans do not collect this information routinely. Claims data do not include it. EHR data might include it and it might be extractable. But the easiest way to get the data is to buy it. And it is available from third-party data aggregators.

These are just a few examples of the potential applications for new forms of personal data that are readily available in the marketplace. The use of these data is governed by privacy laws and ethical standards. My view throughout this book is that these data can be used to engage people more in their healthcare and thereby make them better. I do not endorse the use of the data to improve the business at the expense of the patient.

The use of personal data is not as prevalent in healthcare as in retail or politics. Part of this is due to HIPPA laws on the protection of personal data in healthcare. Part of it is the industry not looking over the fence to what other industries are doing. Part of it is also the medical mind-set of what data are important and what are not.

New Data: Democratization

> *This time around everyone finally appears to understand that real relationships with voters are what drive volunteer engagement in campaigns.*
>
> —Strata Conference, Get the (Data) Vote Out, June 20, 2012

An important aspect of data democratization that materialized in the 2012 political campaigns is the power *over* the data. It has to do with who owns it and how it's used. It's about "open data" and making it available for all to see and use without much filtering or top-down directions. It empowers campaign workers to make their own choices. It facilitates "stake building" that "turns people who are casually interested in a candidate into door-knockers and repeat donors."[33]

A good example of this is the Obama Campaign Dashboard (Figure 4.6 shows a screen shot of the author's dashboard). This was an amazing decision support system for campaign workers. It includes the workers' performance metrics on how they are contributing to the campaign. It provides resources to do their job better, including telephone numbers of undecided voters to call, contact information of other campaign workers, and ways to facilitate fundraising. And it helps keep the spirit and momentum with workers through "get fired up" motivational videos. It communicates the latest about the campaign. And it is fun and engaging.

Source: Obama Presidential Campaign, 2012. Reprinted with permission from the Obama Campaign.

Figure 4.6 Obama Campaign Dashboard.

This availability of data through this state-of-the-art business intelligence application "turns the keys over to the field workers"[34] to enable them to create their own blog, create a fundraising page, or organize house parties or online groups. The results of all the points of engagement are stored in the data warehouse to reinforce a cycle of building information about people's level of interest and activity.

This approach was very successful because it recognized the realities of the workforce for political campaigns. It is composed mostly of underpaid and voluntary workers who believe in the cause and the candidate and pound the pavement to get out the vote. The need to inspire, motivate, and keep up the momentum with these workers such that they were effective with voters was very important. The get-out-the-vote success of the Obama campaign, fueled by data and democratized through its deployment, was a major contributor to winning the election.

In healthcare, there is the obvious need to bring all relevant information about a patient to the desktop of the treating clinician and to do this with compelling clinical decision support features. There is also the need for patients to be much more engaged in the health and healthcare process. A patient dashboard would help. The dashboard would include all relevant data, tools, motivational materials, and decision frameworks. This is not a personal health record or a clinical portal that is often just a depository of clinical facts. It is an action-oriented personal dashboard that includes "my information" and "my journey plans." It is integrated with doctors, insurers, and other caregivers. The most obvious difference between it and EHRs or HIEs is that it is oriented toward people/patients. It is on their terms. It has the express purpose of establishing them as a contributing member of the team.

Another aspect of data democratization is to turn the keys over to people in terms of the types of data they chose to share/provide and the ways they do it. Social media is a major communications transformation for political campaigns. It is an important tool for campaigns to locate and woo supporters and to gather data from them. For voters it is a platform to receive and send information to peers that can advocate or denounce a candidate. The ubiquity and centrality of mobile devices like smartphones complements social media such that the combination might be the dominant communications preference for campaigns going forward.

Two emerging opportunities with social media include sentiment analysis and network power.

Sentiment analysis is a key analytic method for social media. It is a set of methods that detect, measure, report, and exploit attitudes, opinions, and emotions in online, social, and enterprise information sources. It gives explanations about data and associated behaviors picked up in transaction records.

In political campaigns, one can measure sentiment by looking at the number of tweets or likes, as well as whether they are positive or negative.

USA Today partnered with Twitter, Topsy (a data analytic group), The Mellman Group, and Northstar Opinion Research to publish a Twitter "election meter" that tracked the favorability sentiment expressed by Twitter users for President Obama and Mitt Romney, on a scale from 0 to 100 (with 100 being the most positive).[35] For example, on August 27, 2012, the index for the 12 swing states showed a favorability score of 21 for Obama and 41 for Romney. The trend over time is probably more important than the absolute numbers. However, the meter did not correspond with multiple polls that showed Obama ahead in the swing states. One limitation might be that the Tweeters are not representative of the general population, and it is difficult to attribute geography to a Tweet. So it is too early to interpret the predictive power of voter sentiment. It is worth noting that Newt Gingrich had a great positive sentiment index at 84% but lost support among voters and bowed out of the race.

The second opportunity that social media brings to politics is to facilitate political action outside the usual boundaries of campaigns by strengthening individuals' network power. As we have seen, politicians' campaigns might have far more information about voters than voters have about them. In the future, those people who create voter networks of size and significance might have more bargaining power with political candidates than their checkbooks ever could. For example, the amassing of swayed voters could be used as political capital. This use of social media technology might actually transform the playing field such that large contributions are not all that matters and that networks of relationships of trust are what can be counted on to bring influence to politics. For example, the power of networks is understood when one considers that Facebook users have an average of about 400 connections and 50 voters in their network.

In healthcare, social media can extend the truly unique aspect of analytics as developed more fully in political campaigns. As discussed earlier, the special feature of polling is in getting data directly from people and engaging them through the process by participating in the big race on who will be president. We discussed how polling can be applied in healthcare for a moon race for improving well-being and thereby increasing peoples' stake in improving their health and the healthcare for all. But there is one caveat: People might not respond to questions generated in a top-down way.

People comfortable with social media are breaking away from standard ways of giving feedback. For example, rather than responding to a satisfaction survey from the Hilton Hotel, some people prefer to get and give feedback on neutral sites. These sites, like TripAdvisor, provide standard ratings on hotels and restaurants, include star ratings, and perhaps most important, offer qualitative comments from travelers who have been there. These sites are considered more trustworthy, and there is an emerging philosophy that people are taking care of one another with helpful information. (It is also interesting to note that every negative entry this writer has made about a hotel or restaurant has been followed up by the restaurant, which speaks to the power of negative comments affecting other customers and maybe even going viral.) An analogous "HealthAdvisor site" would facilitate the generation of helpful information about the quality and satisfaction of healthcare providers.

People do not necessarily need a structured site like HealthAdvisor to spread comments. Facebook and Twitter are used to inform friends, families, and colleagues about the good, bad, and ugly in their lives. These networks can be very large and can create an epidemic of good or bad will. So it is important to healthcare providers and health plans to learn from the conversations about them, to induce dialogues, and to manage the buzz. What is unusual is that the content is uncontrollable and unstructured. Hence, monitoring sentiment and comments on social media sites is important.

Conclusions

Politics as exemplified by the U.S. presidential campaign of 2012 is the ultimate showcasing of the extraordinary contributions of analytics to the success of an industry. By all accounts, the election was won due to the data-driven execution of the Obama campaign. Analytics contributed to the overall strategies and all the tactics that supported them. The analytics sweet spots include polling, predictive analytics, and data democratization. The potential adaptations for healthcare are profound. On the one hand, polling is an antidote to the relatively blind pursuits of covert big data. It engages the minds of people directly and lays the potential to involve them in the pursuit of their own better health and that of other citizens through better-informed policies. Predictive analytics demonstrated the amazing power of the method to be accurate within a few electoral votes many months before the actual votes were cast, raising the question of whether the voting process is anachronistic. Finally, the influence of social media and mobile, continuing to emerge in politics, lays the foundation for more responsive ways for people to influence politics through the power of networks and to redefine who controls the flow of information and the types of information of most interest to voters and patients.

5

Banking

The big analytics lesson from banking is its understanding of the capabilities and motivations of its customers to carry through on their obligations.

Introduction

No other industry has dug itself into such a deep hole as banking has. Banks are widely viewed as primarily responsible for precipitating the Great Recession due to shaky lending practices related to subprime mortgages. Retail banking customers are dismayed by hidden costs, new fees, the "small print," and the difficulty in getting a loan. The public's confidence in the industry has been shattered...worldwide demonstrations have accused the industry of greed and corruption. Politicians from around the globe have responded with extensive regulatory requirements.

The business climate for banking is dark. The *Wall Street Journal* reported on the "grim new reality" for the industry characterized by "slow economic growth that tamps down loan demand, low interest rates that pressure investment returns, volatile markets that inhibit risk-taking and tighter regulation that adds to bulging costs."[1] Finally, McKinsey & Company concluded that "without radical action to shrink balance sheets, cut costs, and increase revenues, banks will be unable to attract sufficient new capital from the investment community and to play their critical role in underpinning economic recovery and growth."[2]

So what's an industry to do? How can the resolution of this crisis, driven by tremendous pressures to stabilize and grow the business, deepen its capabilities in analytics? What are the unique analytic sweet spots of the industry? And how can the sweet spots be translated and adapted to healthcare? The bottom line is that it comes down to two solutions with the acronyms HICS (Health Improvement Capability Score) and AMT (All Multichannels Together). This chapter provides the guided tour to understand the industry and its associated analytics and what it means for healthcare.

First, let's try to understand the industry. For most of us, banking appears to be very complicated. For example, does any layperson really understand what happened when subprime mortgages were split up into multiple securities that Wall Street packaged into investment products? Even investors seemed clueless about the risk associated with these securities and then lost their collective shirts as millions of customers were unable to make payments on mortgages that they should never have been given in the first place, and subsequently defaulted on. And the whole house of financial cards and worldwide economies collapsed. In the unforgettable words of Gordon Gekko in the movie *Wall Street*, "Greed, for lack of a better word, is good." That is, unless it makes people do foolish things and goes too far.

At its core, however, banking is really quite straightforward when compared to other industries. Simply put, banking makes money off of money. It needs to secure funds either through its banking customers' deposits and other money holdings or through other institutions that lend it money (such as commercial banks). Then it turns around and lends money to others and makes profits on the spread between what it paid for the funds and what it can get in return for lending it. It has a relatively narrow range of products to offer its customers, including checking accounts, saving accounts, CDs, mortgages, credit cards, trust services, and perhaps brokerage services. It is narrow when compared to the broad range of products offered by retailers. It is simple when compared to supply-chain issues in manufacturing or the complexity of what constitutes success as in healthcare. There are many performance metrics in banking, but it mostly boils down to using money effectively as measured by returns—return on investment (ROI), return on equity (ROE), return on assets (ROA), and so forth. Compare this to the range of diverse measures in healthcare spanning clinical outcomes, adherence to guidelines, economic outputs, costs, and much more.

The American banking system is large. It consists of 7,513 FDIC-insured banks and savings associations owning 87,873 bank offices and 10,319 savings association offices. Deposits in FDIC-insured organizations totaled $8.85 trillion in June 2011.

Additionally, in 2011, there were 7,339 credit unions with assets totaling $914.5 billion.[3] Most of the banks are community banks but most of the money is held in larger banks.

Banks have a long history going back to 9000 B.C., starting by making grain loans to farmers and traders. In more modern times, banks became a fixture in almost every community because they provided services that every community needed, just like local schools, fire and police protection, and doctors and hospitals. People went to their local banker to start a business, buy a house or a car, send their children to college, and provide a "monetary home" for their life savings. Bankers were leaders in communities and gave back to help the communities prosper, not just through the money services they provided but also with dedication to the needs of the community in supporting charities and important community initiatives. The 1946 movie *It's a Wonderful Life*, directed by Frank Capra, captures the (sentimental) essence of this community obligation. People knew and trusted their banker just as they did their doctor. The doctor took care of their health welfare and the banker took care of their economic welfare.

But the banking experience that most of us have today is quite different. Rather than walking into a bank office and exchanging greetings with the tellers and the manager, the more typical communication now is putting an ATM card into the slot and "greeting" it with a punched-in password. Ditto for Internet banking. Of course, healthcare has changed as well. Doctors do not make house calls anymore and have been "industrialized" into medical groups that produce efficient eight-minute slots for patient consultations with great efficiency in provider-centric complexes.

Nostalgia is warm and fuzzy and might not be relevant for today's solutions. Community has given way to "what's best for me." Consumers want the most convenient and frictionless transactions. Social media communications are the rage but are often more akin to stalking people and extracting information than to building deep relationships. Efficiency has trumped relationships.

So banking has two faces. On the one hand, it was created out of a strong commitment and connection to its primary customer, the people in its communities. On the other, its basic formula to "make money off money" requires a laser focus on the returns on A, E, and I to satisfy its superordinate customer, the investors, in a very turbulent and competitive environment. The balance between these two customers got out of whack, with the former becoming subservient to the needs of the latter. The resultant depersonalization of much of banking today and its associated business failure consequences might be a necessary "hitting bottom," as with a person overtaken with alcoholism, to motivate it with a sustained will to reinvent itself. Appreciating and servicing customers as a back-to-the-future success strategy might be the hardest, most counterintuitive, and surest way to achieve growth.

The same can be said for the health insurance industry. Its "primary" customers are its members: those who sign up, pay premiums, and receive payment coverage for the care needed to treat illnesses. But the more important customer became employers who actually paid (most of) the premiums out of employee compensation and called the shots about what benefits are provided and what health plan(s) are made available to employees. As employers became more concerned about their escalating insurance costs, they tended to shift costs to employees. For example, many employers offered/required employees to enroll in high-deductible/health savings account plans, which was not necessarily done for the health welfare of employees. Insurance companies were happy to satisfy employers with multiple products and different price points and got out of the business of managing medical costs and real member needs. Members were regarded as a cost to be managed and were often treated by customer service representatives as irritants that needed to comply with business process requirements.

Health insurers became the lowest-rated industry in terms of customer engagement.[4] (Forrester Research does surveys on customer engagement. The health insurance industry had the lowest average score, at 51%. The average score for banking was 66%. The highest-scoring industries were retail and hotels, at over 80%.) Many business practices of insurers were considered unfair, such as denials for care associated with preexisting conditions, and were subsequently regulated through Obamacare. Other regulations of Obamacare required that insurers become more transparent on the quality and costs of the benefits provided and that they compete in health insurance exchanges to offer their products directly to a potentially large segment of Americans. Additionally, market-driven changes in the way providers are paid—that is, from fee-for-service to global payments—and in the preference for integrated and coordinated care that is measured by outcomes and not by the process of the care, are also changing the perspectives on "customer needs." And what member customers want will be quite different from what employer customers want.

So the two industries, banking and health insurance, have a lot in common. Both have reached bottom with their customers. Both have been the subject of extensive political and public debates that have spotlighted weaknesses and excesses. Both have been slapped with extensive and onerous regulations. And both need to get back to the future with strategies and practices that give people a reason to do business with them.

Further, they are both increasingly commoditized industries, for example, processing claims and money transactions and issuing insurance or loans, and both need to get a new raison d'être to prosper with profitable business lines. For example, for one major insurer, more than one-third of its $34 billion in revenue comes from noninsurance revenue streams. Its CEO said, "We're evolving into a 'Health IT' company with a health insurance component."[5] For other insurers, they are evolving to become a full-service healthcare company. Cigna has ramped up its customer outreach with 24/7 customer service, mobile applications, social media, decision support tools, and access to health coaches. It is also expanding its engagement with accountable care organizations and partnering with physicians and hospitals more closely.[6] But many insurers, perhaps most, dismiss these radical solutions as overkill and think this is basically a public relations problem that can be solved by rebranding and customer communications about how much they care.

Banking knows it must change and agrees with Rahm Emanuel's immortal words to not let a good crisis go to waste. The industry also understands the "what" more than the "how" in the "back-to-the-future" customer strategy. Many believe that the reinvention and rebirth will be driven by data. When a panel at the 2012 Strata conference on the future of banking was asked, "How pervasive is data analytics and will it provide the competitive edge to transform?" Richie Prager from Blackrock said, "It's a huge edge." But Allen Weinberg from McKinsey said that the state of the art of analytics is quite different from the state of the practice.[7]

And this is at the core of the sweet spot and potential lessons learned from the analytics of banking. Can an industry under more stress than any other (be forced to) make use of analytics in ways to reinvent itself? If not, and if it wastes the crisis, why would any other industry find the guts/purpose/commitment to adopt the best that analytics has to offer? And if it does, how do the analytics mature in unpredictable and totally productive ways?

Industry Challenges

McKinsey's review of the banking industry in late 2012 painted a dismal picture of its present and future state.[8] Revenue growth is down to 3% (compared to 9% in 2007). There were no improvements in costs and there was a deterioration in capital efficiency. This led to a decrease in profits of 2% from 2010 to 2011 and a 15% drop compared to 2007. Global average return on equity (essentially net income after taxes) fell to 7.6% in 2011 (with the U.S. at 7%), down from 13.6% in 2007, which is half its peak value before the financial crisis. The industry as a whole is far from earning its cost of equity, which would require an ROE of 12%. If it does not reach its cost of equity, investors will go elsewhere.

The drivers of these numbers are many. The first is that there are fewer customers demanding loans. Consumers are shying away from debt, paying down balances, and using credit cards less since the boom times ended in 2007. This is seen as a long-term consumer trend toward increased savings, lower consumption, and lower debt. So, paradoxically, there is more funding available for loans through increased savings but poorer use of capital because it is not optimizing returns through profitable lending. In addition to less lending demand, banks are providing less supply of loans because of stricter guidelines about taking on credit risk.

To compensate for this loss in lending revenues and the increased burden of regulations that lead to higher costs (discussed later), banks have looked to new fees to generate income. In fact, one bank official told me that one of the most profitable customer segments consists of people who chalk up lots of fees, such as overdraft fees. Again, paradoxically, these people are more likely to need loans and to provide hugely profitable fees but are not the best risks for loans. However, when banks, most notably Bank of America, announced that they would institute new monthly fees for debit cards, customers left in droves and the backlash caused the banks to reverse the fees. And how about those $39 late-payment fees for credit cards? These are also being tamped down by federal action (see the later discussion).

The pipeline for loans is also eroding because the "shadow banking" system is siphoning off regular banking customers. These shadow companies, including retailers, automobile manufacturers, stock brokers, insurance companies, and others, offer a full spectrum of bank services, including loans and mortgages, credit cards, and money market accounts with checking account features. For example, Walmart has become a financial service giant, offering "Money Centers" within many of its 1,800 U.S. stores where customers can pay bills and cash checks.[9] E-trade and Charles Schwab offer free checking accounts with no ATM fees and the convenience to combine cash with stocks in a one-stop-shopping portfolio.

In addition to going after typical banking customers, there is a growing activity from nonbanking entrepreneurs, primarily from the retail industry, to capture the "underbanked" population. This population might not have credit cards or might be perceived as being too risky for traditional loans. For example, Walmart and American Express provide financial services through a prepaid debit card aimed primarily at low-income customers.[10] Start-ups also offer prepaid cards. For example, PayNearMe caters to customers without a credit card with the corporate tag line "You can now borrow online and pay offline with cash."[11] Green Dot is a leading provider of reloadable prepaid cards that can be used wherever traditional cards like MasterCard and Visa are used. Their tag line is "Big Banks, No Thanks."[12] And NetSpend Holdings promises "The better way to bank" with prepaid MasterCard and Visa cards.[13] In addition to earning revenues from banklike products, these companies are tapping another critical "natural" resource that is highly valued in the retail industry; that is, vast data on how customers are spending their money.[14] One of the most important uses of these extensive vaults of data on purchasing patterns and a wide range of personal data is to refine credit risk determinations that can open the funnel on new customer segments for loans. Of course, these data can be used for pinpoint marketing as well. (Both of these issues are discussed in detail later.)

To compound the disruption, other groups are questioning the need for intermediaries like banks at all, especially when plentiful information on buyers and sellers is transparent and risk can be determined forthrightly. These groups facilitate direct peer-to-peer lending through lending clubs, between friends and families, and through crowdsourcing. For example, Prosper.com "connects people who want to invest money with people who want to borrow money, offering "seasoned returns" to investors of 9.69%.[15]

Last, but not least, new regulations are a big challenge for the banking industry in terms of the cost associated with compliance and the risks associated with not anticipating regulators' needs. The Dodd-Frank Wall Street Reform and Consumer Protection Act was signed by President Obama in July 2010. This followed European finance ministers' approval of similar regulations for much of Europe a few months earlier. However, the Dodd-Frank bill has not yet been fully defined (as of the end of 2012) in terms of the specific requirements, causing banks to be ever vigilant to the potential need for requirements. In general, the bill includes broad powers for the government to monitor systemic financial risk and to intervene where it deems necessary. Banks will be required to hold much higher levels of equity capital, and their ability to earn profits will be restrained by changes in the regulatory environment. In addition, commercial banks' ability to take investment risks will be restrained. And the government can require banks to get smaller and even to liquidate if the bank is posing financial risks and approaching bankruptcy.

An example of the management challenges in achieving compliance is described by Julie Wooding, a consultant at FICO.[16] In the example, a CEO says the new formula required by Dodd-Frank and Basel III for calculating deposit-insurance premiums takes 18 employees working on four committees to gather 60 different numbers, which are plugged into 20 separate equations to derive the bank's "performance score." Later we will discuss how these new work processes can be managed in the most efficient way through perfected enterprise analytics.

Another regulatory concern for banks is the creation of a Bureau of Consumer Financial Protection that will oversee and regulate the issuance of mortgages, credit cards, personal loans, and retirement plans. The bureau has a $500 million budget that does not require Congressional approval. The new agency has the authority to enforce its rules on banks with more than $10 million in assets. As of the end of 2012, the bureau was also still a work in progress. In general, banks will be required to maintain much larger levels of capital. Credit card issuers will be required to operate under modified rules that are much more consumer-friendly than in the past. Debit card processors will be required to charge lower fees on transactions.[17]

Key Business Drivers and Strengths

In all industries, the key business approaches to improved profitability are to cut costs, improve the balance sheet, and increase revenues. These are heightened in the banking industry because of all the challenges discussed, which force a renewed effort to make the business work much better.

Cut Costs

Improving banking productivity through cost cutting is necessary but not sufficient to achieve the ROE benchmark. McKinsey notes that banks would have to reduce their expenses by up to 6% a year between now and 2015 to achieve it. But, according to their survey data, only 2.5% of banks achieved annual cost reductions of 4% or more in the years from 2000 to 2010.[18] Clearly, there might be greater routes to profitability through revenue enhancement; there are, however, clear steps for improving efficiency, and these have to do with changing infrastructure and optimizing channels.

The first infrastructure issue is a *move away from bricks and mortar* (bank offices and branches) for the delivery of many banking functions. Indeed, almost all typical transactions could be handled digitally[19] through online channels that include self-service approaches. Branches are very expensive and are used less and less for usual retail banking transactions. The branches could be repurposed for more revenue-producing purposes. For example, some have suggested an ultra-makeover, to a cool and friendly Apple store–like place where customers can come and experience new ways to manage their money in a comfortable, low-pressure environment. Financial coaches might suggest ways to satisfy their personal needs by considering compelling financial products from the bank.

This movement away from the physical space should also happen in healthcare. I do not look forward to my visits to the medical megaplexes where much care is delivered. Clearly, some "hospitality" functions of hospitals have improved, including pianos in the lobby, single rooms, and better menus. And certainly very sick patients need to be treated in environments that aggregate resources, such as the many specialists who seem to be needed. But most care is pretty routine and can be handled virtually, which would please a large segment of customers. But it seems that one has to have an appointment for almost anything, and this can be very inefficient for the customer. Perhaps the physical space requirement exists more to satisfy the fee-for-service system and the self-serving doctrine that it is better to be safe than sorry than to get the job done on the customers' terms. More on this in a bit in the "Strengthening the Multichannel Platform" section.

The second infrastructure issue is the need *to modernize IT systems* to meet regulatory demand. If all regulations in the pipeline were applied with immediate effect, the top-13 banking global players' ROE would fall from 20% to 7% due to the increased effort required to comply.[20] After the regulations are known, or predicted, many of these can be automated to dramatically reduce operating costs. An example of the complexity of regulatory demands is in the area of rules on money laundering. Simple tasks like checking the names of clients against those on a sanctions blacklist can be very complicated for banks, which might have thousands of customers with the same names as those on the blacklist. Each becomes a false positive that could embarrass the bank and ruin a client relationship. So banks turn to computers that can aggregate data from a variety of sources, including the customer's nationality and address, the names of family members, and whether they have travelled to or received money from countries on sanctions lists.[21] Additionally, upgrading legacy systems and applying advanced analytics could cut the time for many business processes; for example, cutting the cycle time for a mortgage decision to minutes rather than days.

A valuable byproduct of upgrading IT infrastructure for regulatory compliance and improved business processes is the use of new data and system capabilities to provide new insights for other banking functions such as serving customers better. This comes from full integration of all data from its various business lines and channels. Indeed, the upgrade can enable a full digitization of the business. One way of picturing how digitization can help customer service is the ability to make decisions in real time with decision rules and machine learning that lead to the right offer, to the right customer, at the right time.[22]

The third infrastructure issue is full enablement of *multichannel* capability, especially to "go mobile" and "go social."[23] Mobile becomes the new hub of transactions and communications. It is the channel preference for consumers in many industries, supplanting more fixed devices such as computers, land lines, call centers, and (soon to be archaic) ATMs across town. It provides new ways for paying and borrowing "on the go." And most important, it is totally available to the customer to receive information and messages at the right time.

Mobile banking might not be profitable for usual banking transactions, but it is "sticky" in terms of retaining customers largely because it satisfies customers' need for convenience, which is very important. And it might become the new table stakes on providing banking services. However, it could be a powerful platform for creating new revenues if banks turn their focus to achieving good financial outcomes for customers, thereby broadening the scope of services provided.

Related to mobile is social media in terms of capitalizing on customer preferences for peer communications. In one way social media is a way to create "mass" intimacy in a digital age. For example, social media can be used to elicit feedback about bank initiatives and to spread messages. In another way, social media is a surveillance tool to monitor news channels, blogs, forums, Facebook, and Twitter, to monitor whether sentiment is positive or negative and to then counter the bad and reinforce the good. Social-media data can also be used to understand customers' characteristics and predilections for supplementing marketing data by mining personal data.

Finally, a common way to attempt to reduce costs is through *mergers and acquisitions*. The regulatory burdens are especially difficult for small banks, and an M&A might be the surest route to (a form of) survival. Other reasons espoused for mergers and acquisitions include other ways to consolidate expense reduction, increase market power, reduce earnings volatility, and improve scale and scope economies. However, there is no compelling evidence to support the proposition that mergers and acquisitions provide a gain in value or performance, although acquired firm shareholders tend to gain a lot at the expense of the acquiring firm.[24]

Improve the Balance Sheet

Banks need to work with less capital and use what they have more efficiently. According to the McKinsey report, banks have made significant efforts to stabilize their balance sheets. For example, the industry Tier 1* ratios increased 11.7% in 2011, compared to 8.2% in 2007. Tier 1 capital has increased sharply 57% since 2007.[25] The percentage of risky assets to total assets declined by one percentage point. These improvements show relatively good progress, largely spurred by government requirements.

In order to do more with less capital, banks are working on *improving credit performance*; that is, reducing the number and impact of underperforming loans and reducing credit costs. Assessing credit worthiness is at the core of the analytics that banking usually does very well. But it was stunned by the subprime mortgage crisis that neglected the core rating principles of understanding credit histories and the ability to repay. To be fair, much of the subprime mortgage activity was not done by traditional banks but by nonbank independent mortgage originators such as Countrywide. These "shadow banks" were able to mask credit issues from investors and regulators through the use of complex, off-balance sheet derivatives and securitizations.[26] In fact, about 50% of the growth in mortgages in the early 2000s was attributed to nonbanks.[27] This is a case of the few bad apples causing the whole bushel to rotten. However, the result, growing out of the crisis, is a refined approach to establishing credit risk. We see this as the unique banking sweet spot and will discuss it shortly.

Grow the Top Line

The biggest opportunity for banks to improve their profitability *might* be to become *much* more *customer-centric*. McKinsey states that a customer-centric view leads to greater loyalty, higher cross-selling, less attrition, and ultimately higher sales and profits.[28] It is largely a marketing and sales function and includes cross-selling and up-selling existing customers, attracting new customers, and opening new markets for even more customer segments. The strategy, well known in the

* Regulators use the Tier 1 capital ratio to grade a firm's capital adequacy as one of the following rankings: well-capitalized, adequately capitalized, undercapitalized, significantly undercapitalized, and critically undercapitalized. A firm must have a Tier 1 capital ratio of 6% or greater, and not pay any dividends or distributions that would affect its capital, to be classified as well-capitalized. Firms that are ranked undercapitalized or below are prohibited from paying any dividends or management fees. In addition, they are required to file a capital restoration plan.

retail industry, is that by knowing customers better, with ever-increasing breadth and volume of data, and then segmenting them into useful marketing clusters, the retailer can pitch the right product, at the right time, through the right media to the customers and increase sales. In the retail chapter, we discussed how Target homes in on women in their second trimester of pregnancy and uses this "opportunity for habit change" to sell a host of products related to baby and family.

Additionally, this type of targeted sales based on segmentation can indicate those individuals who can be most profitable through a high-touch approach and those who are less profitable and can be managed adequately with high-tech approaches. The potential for granular microsegmentation to approach the realization of a virtual "market of one" can improve relationships and offers in other ways as well.

Wells Fargo Bank is known for its commitment to customers. Its vision statement reads, "We want to satisfy all our customers' financial needs and help them succeed financially." It also states, "Our vision has nothing to do with transactions, pushing products or getting bigger for the sake of bigness. It's about building lifelong relationships one customer at a time."[29]

In healthcare, health plans are changing their tune with customers. It's not just about instructions to members on how to submit claims. For example, the tag line for Tufts Health Plan is "No one does more to keep you healthy." WellPoint's mission is to "improve the lives of the people we serve and the health of our communities." Kaiser Permanente focuses on "Thrive...Find your groove...Wake up smiling." The question is whether the new tune will lead to good customer experiences and outcomes and better business outcomes.

There is a good deal of vendor hype on how big data will help marketers. For example, one vendor says that big data will provide the "opportunity to understand customers at a depth that harkens back to the past of neighborhood stores and local banks" and to develop "deep, relevant and constantly improving dialogue with these customers"—and to do so at an immense scale. This proponent provides an example of pitching a product at the right time. For example, the bank knows that a customer frequently accesses her checking account via her mobile device; she also uses Web banking and handles all her banking communications electronically. Put this information together with the fact that she most often accesses her accounts in the evening at 8:00 p.m., and the data might suggest that the offer would best be sent via e-mail or text at 8:05 p.m.[30] But, you know, I am not sure it matters when and how an offer is presented for a new and improved checking account. The e-mail might go directly to the trash bin if the recipient knows that it comes from the bank.

Another example comes from Singapore. Citigroup keeps an eye on customers' credit card transactions for opportunities to offer them discounts in stores and restaurants. (Say what, from a bank?) The customer swipes a credit card; the analytics checks the time of day, the location, and previous shopping or eating habits and then recommends, through a text message, an Italian restaurant just down the street that meets the customer's profile. It could result in the bank getting a second card transaction and a cut from the restaurant. The model for this is Amazon's online store, which recommends items that a customer might like based on previous purchases and on what similar customers have bought.[31]

Another approach to knowing the customer well is to go beyond product pitching and restaurant referrals and give them useful information products. For example, Britain's Lloyds Banking Group wants to go beyond answering "what's left in my balance" to advise customers on a prediction of how much they will have available after all their usual bills are paid. And products like Quicken and Yodlee sell software to banks that they can provide to customers to help them manage their money.[32]

Finally, new revenues can be grown by repurposing the branch office as alluded to earlier. An example of this is OCBC, which "aims to convey honesty, sincerity, and simplicity," and according to them "may be the coolest bank Gen-Y has ever seen." They use retail stores modeled after a shopping experience and create credit cards with customized designs to express their personality. Also, SNS Bank has storelike outlets, tablets that customers can use, and extended hours.

Quite frankly, however, financial products like checking accounts and CDs are pretty stodgy and do not have the pizazz of iPads or the broad array of the latest and sexiest products from Target and Apple. And I am not sure how many profitable customers would be attracted to the bank by having their dog's picture on their credit card. Certainly, high-end retailers attempt to turn their showrooms into a living room, including a piano player, to make the customer comfortable and also primed for buying. And wealth management services from banks can have very tony physical environments for its richest customers. The question is whether this radical bricks-and-mortar reinvention will attract more than this upper slice of customers, and perhaps the cool Gen-Y'ers, and deliver more revenues. This transformation would require much more than a reuse of physical assets as well. It would also require the staffing of branches to change from bank tellers to financial friends and advisors.

Culture intrudes. We have talked a lot about possibilities for growth so far and some concrete examples. But, as mentioned at the outset of this chapter, the state of the art in banking is quite different from the state of the practice. There are examples of best practice, for sure, but for many banks the journey to sustainable growth has just begun and it has more to do with the sociology than the technology. Sean Rowles, EVP for Citizens Bank, formulates a "banking hierarchy of needs," ranging from the base of the pyramid on what he calls survival mode to apathy mode in the middle and to advocacy mode at the apex. Many banks are in survival mode and focus on regaining trust and being safe in terms of taking on risk and complying with regulations. But this strategy does not lead to more revenues and might actually increase costs. The second level is apathy and is marked by being convenient, as with multiple channels, and by being competent in the usual banking initiatives. This can produce modest increases in revenues. At the top of the pyramid is customer advocacy where the customer is king and queen and the approach is to "meet my needs, be approachable, and know me."[33] Clearly, a bank's success in achieving sustainable growth will be how fast and well it can get up the curve.

Analytics Sweet Spots

Relative to the key business drivers just described, this section turns to selected analytic sweet spots that support a few business solutions that are unique to banking and offer promise for healthcare (see Figure 5.1).

It boils down to two sweet spots: (1) infrastructure: transactions processing and (2) predictions: risk.

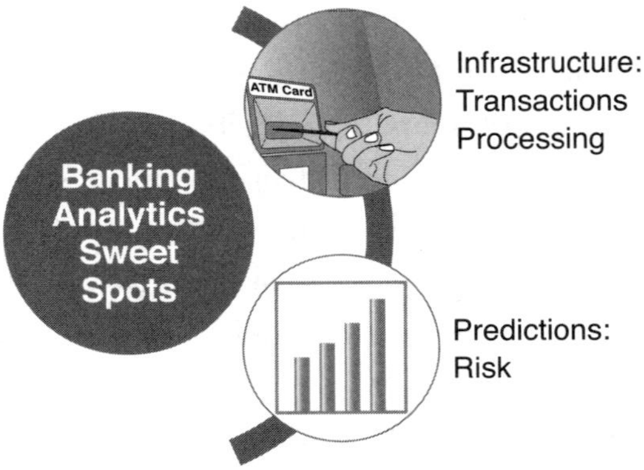

Figure 5.1 Banking analytics sweet spots.

Infrastructure: Transactions and Processing

The banking industry has been a pioneer in new technologies for handling transactions. It started with ATMs, automated teller machines. ATMs were a significant advance in service and convenience for customers and were no small miracle of technology. No longer did customers have to enter the bank, stand in line, and wait for human tellers to cash checks, receive cash, and make deposits. The success of the self-service approach was picked up by other industries, most notably energy, in which people could pump and pay for their own gas at stations. Today, people can get medicines and DVDs and much more from self-service machines.

The first ATM was put into use over 50 years ago in a shopping center in Ohio. Today, ATMs are everywhere. In fact, there are 370,000 automated teller machines across the United States, or about 1 for every 296 people. Compare this to the number of doctors per person in the U.S. (2.3/1000)[34] and you can see that people "demand" about 50% more ATMs than MDs. Perhaps if the provision of medical care were more virtual and as efficient as banking with ATMs, the demand would increase, prices would fall, and outcomes would improve.

The driving force for ATMs was most certainly cost reduction. But consumers grew to love them even though they harbored worrisome concerns about being robbed at the machine. Similarly, would anybody in their right mind want to handle extremely flammable liquids to save a few cents on a gallon of gasoline? People were willing to accept these risks in order to lower the friction costs of being served by slower humans. And beyond that, the industry developed the "new normal" through mere force of presence, convenience, and rewards.

ATM usage is on the decline. The preferred banking method now is Internet banking, with more than 50% of customers using it. But it too is being eclipsed. Just as ATMs reduced the demand for branch office visits and the Internet supplanted the need for field trips to the ATM, mobile is clearly the emerging channel of choice today. Mobile combines higher levels of convenience with new paperless technologies. For example, mobile apps allow one to take a picture of a check with a smartphone and to deposit it on the fly.

Banks are also looking to technology to improve customer (self) service through natural language processing. Citigroup is working with IBM on how to advance "customer interactions" by using IBM's supercomputer "Watson."[35] Recall that Watson was the computer contestant that beat the best-performing humans at *Jeopardy*. The Watson system is powered by its DeepQA software, which is able to ingest large quantities of information and respond to questions with a series of options and related probabilities. For example, it could "know" the customer's financial situation thoroughly and be able to answer customer questions as well as or better than humans and to offer financial advice and products. It is not clear how well robots will contribute to improving customer centricity, although they certainly will reduce costs.

Healthcare is the other industry where Watson is looking for opportunities. IBM is working with WellPoint and Cedars Sinai Cancer Institute to train a new generation of robotic physician assistants. These could inform doctors of the pertinent points on the vast medical research on cancer, as well as a patient's medical and genomic history, to help with differential diagnosis.[36]

What banks need now are A*M*Ts, or All Multichannels Together, to allow for seamless transactions through many channels, especially with mobile at the hub. With all channels integrated, the amount of information about a customer multiplies and can be harvested for other business needs. However, the information system infrastructure needed to support the multichannel environment and to wring out insights from all the emitted data is more complex and costly than most existing core systems.

Similarly, banks are under great pressure to anticipate and respond to regulatory requirements. The risks of noncompliance are huge. And as they venture up the curve to be truly customer-centric, they will need to gather and integrate all the information from a variety of sources to know customers better in order to offer tailored products and keep them away from savvy upstart competitors.

To master all these demands, banks need to perfect two fundamental pieces of analytics: Get the business basics right and strengthen the platform for the digital future.

Getting the Business Basics Right through Analytics

Banking needs to make the business work better. It requires a commitment to industrialize its processes. This involves improving its core administrative IT systems and, perhaps, modernizing legacy systems. It has to do the basics right and then reap the rewards in efficiency gains.

Perhaps the most fundamental improvement is to organize the data around the customer. Banks, like other industries, operated in silos related to business units, product teams, or acquisitions, and the information systems were centered on each business's needs. There was little coordination among the different groups with the whole enterprise in mind, and customers might have received duplicative offers. The enterprise was simply not able to get a full picture of its customers across the various parts of the business.

What does this full picture, or 360-degree view of the customer, provide? It can inform the business about (1) the products the customer uses and her interactions with customer service, (2) how she responds to incentives and conducts banking business, (3) what she buys from other banks, and (4) how she uses different channels.

Now, this vision of a 360 view of the customer has been discussed for decades. A number of banks have taken the leap, but most have chosen not to. After all, it is a big and expensive deal and usually is associated with large-scale legacy system improvements. The ROI for a 360-degree view is more a vision than an evidence-based assurance. Indeed, a lot of the push for a customer-centric view and its associated IT requirements comes from vendors marketing the "art of the possible."

An example is a report from IDC Financial Insights on "Banking for Success: Using Analytics to Grow Wallet Share."[37] IDC is a provider of research-based advisory and consulting services for the finance industry. The white paper advocates the value of customer relationships and the analytics that support it (customer analytics) to increase profits. Its primary business case is that customer analytics leads to improved marketing success. It gives a scenario, based on client data, of the potential benefits. The scenario is for a retail bank with ten million customers and $2.75 billion in revenue. The total estimated benefits amount to $43 million from the following categories:

- Reduce customer churn $21M
- Increase revenue per customer $16M
- Increase revenue from loan activity $5M
- Reduce cost of marketing $1M

The benefits, or $43 million, represent about 1.5% of revenues. But the cost of producing the benefits is not included such that an ROI cannot be calculated. Upgrading a legacy platform can easily take tens of millions of dollars. So if this is the state of the art in making the case for customer analytics to improve marketing results, it is incomplete. And it is understandable why there might be a reticence to invest in it.

The other fly in the ointment about moving to a 360 view might have more to do with the sociology than the technology. The sociology has to do with the reservations that banks might harbor about being fully customer-centric. They might sense that customers will not be receptive for a variety of reasons; for example, the trust issue, the limited portfolio of products, and the culture of moving from tellers to advisors.

While the sociology ripens and the evidence accumulates on the eventual need for moving toward a customer-centric platform, there are other analytics that demand quick action. The biggest one is that regulations are overbearing and modernizing core systems to comply is a survival issue. Legacy systems might not be flexible enough to adapt well to the changes. For example, rules might be hard coded into legacy systems, requiring extensive software modifications. So the need for flexibility and scalability often spells the need to retool or replace legacy systems. The good news is that software and hardware are getting much better to accommodate real-time decision making, including the use of in-memory databases to work with large sets of data with minimal time delays, cloud computing, service-oriented architecture, and capacity to manage big data from old and new sources such as unstructured data and social media.

A healthcare example of the benefits of modernizing core administrative systems comes from the IBM publication "The Value of Building Sustainable Health Systems."[38] This white paper offers quantitative estimates of the benefits for health plans in adopting a tiered transformation road map based on analytics development and related competencies. The first tier of the model is to improve core administrative systems, and the second is to address proactive regulatory management. Later stages include innovations in market offerings and business transformation and agility. The research estimates that with a "fully deployed" analytics program as specified, a mid-size health plan could net a potential $644 million annually in economic benefits.

The first stage involves simplifying IT systems, data, and infrastructure, as well as modernizing claims, benefits, and enrollment processes. The estimated annual benefits are $129 million, which is derived mostly by reducing IT costs and harvesting the data to reduce medical costs. In the second stage of regulatory management, most of the overall benefit of $61 million comes from reducing the administrative cost of compliance and by subsequently using the new data capability to reduce medical costs.

Again, as with the IDC example, the investment cost to achieve the benefits is not stated. Also, the benefits represent capturing the fully matured opportunity, that is, what is possible if everything is accomplished and if all the stages in the implementation proceed without a snag.

Strengthening the Multichannel Platform

In addition to strengthening the basics, banks must build new capacity in the areas of mobile and social media. Mobile is the answer to a strong customer demand for convenience, but it also provides business opportunities for marketing and strengthening customer loyalty. And social media is a consumer force not to be ignored. It is an opportunity area to build customer intimacy by being socially engaging. And it can aid marketing and mitigate reputation risks.

Today, many banks have projects underway to pull together customer data from all channels—branch, ATM, online banking, mobile banking, call center, social media sites—in one place as an "intelligent multichannel bank." The payoff is to mine the data in real time and use it to cross-sell, up-sell, detect fraud, and keep customers in the right products.[39] Additionally, the multichannel integration produces new data to facilitate pervasive analytics as with predictive modeling and including microsegmentation. And the platform can make the data available in near real time to service all aspects of the business.

In healthcare, the emphasis and excitement are less on building and mining multichannelintelligence and more on using mobile health (mHealth) applications to improve care. The number of health-centric apps on Apple devices has more than quadrupled since 2010 to about 19,000, with more than a quarter for health professionals.[40] The doctor's black bag has given way to digital tools. Smartphones have amazing capabilities, including the camera, GPS tracking, and call history, which can help with medical treatments. And social media is becoming an important channel of communication, especially with younger patients, such that the phrase that the "doctor will see you now" is supplanted with "the doctor will Tweet you now." The GPS feature of smartphones has enabled interesting applications for monitoring patient progress. For example, pain and depression can result in less activity. An app that tracks how often a patient sends text messages and places calls, and how often he moves and where he goes, when compared to previous behavior, can provide an early indicator that the patient is withdrawing or becoming less active and could indicate a flare-up of illness.[41] Similarly, the camera feature can be used to take pictures of skin lesions that might be indicative of melanoma and send it for analysis and advice about whether to consult a dermatologist.[42] And because iPhones are what teenagers live and die for, the perfect channel to send health messages is through Twitter, including hyperlinks to cool Web sites with teenager-friendly materials, rather than the usual stodgy medical handouts.[43]

Predictions: Risk

The cashier swipes your card and smiles. Before she hands it back, complex mathematical models hundreds of miles away crunch the data: If you've hit your credit limit, should it be extended? Is it really you using the card?[44]

We are all familiar with filling out the forms to get a credit card, car loan, and mortgage. We know the importance of our FICO score, which is the primary metric that determines whether we will get a loan and at a good rate. We also know about how the lending threshold for home loans was lowered such that subprime loans were issued on a mass scale, which contributed mightily to the meltdown. Subsequently, capital dried up and one had to have a stratospheric FICO number to get a loan. All this, and disruptive innovators entering the space, spurred a refinement in thinking about how to assess the likelihood that a customer will follow through with the obligation to repay a loan.

What is FICO and why is it so important to banking...and to healthcare? FICO is an acronym for Fair Isaac Corporation, the creators of the score. (One might have imagined that the acronym would have stood for something really momentous, like the Financial Index that Controls Outcomes.) It is a summary score of an individual's past performance in paying off debt and a prediction of future behavior to make good on a loan. It is derived from data on payment history, current level of indebtedness, types of credit used, length of credit history, and new credit. People receive scores between 300 and 850. Above 650 indicates a very good history. FICO is usually the most important indicator of the person's credit trustworthiness to follow through and pay off a loan. Often it is the only number a lender really looks at.

In healthcare, there is no clear analogue to a FICO score. But there should be. We propose the HICS, the Health Improvement Capabilities Score. This would be an overall indicator of an individual's capability to do their part in getting better. The HICS value would appear on each patient's chart and would guide treatment decisions just like diagnostic values from laboratory tests do. It is very important because much of the untapped resource for health improvement lies with the individual. In fact, as discussed previously, healthcare accounts for only 10% of improvements in health functioning, whereas personal behavior accounts for 40%.

Doctors lament the fact that patients do not "comply" with "prescribed" medications or treatment "orders." Yet there are many reasons why the HICS might be low, and caregivers should be fully cognizant of them if their goal is improved patient outcomes. More on this later after we learn more about the specifics that banking can teach us about making judgments on capabilities.

What's clear is that lending practices are changing and a broad swath of new data is being used in predictive modeling to understand risk better. The first response to the crisis might have been to not lend because of fear of loan defaults, which were high on the list of oversight metrics. Some banks reasoned that they should just wait until the smoke clears. On the other hand, if banks do not lend, new competitors outside the banking industry are more than willing to step into the gap and gain the business.

So the reality is that banks must lend, but must do it differently. In an address to the FICO World 2012 Conference, PNC Executive Vice President and Credit Executive Gordon Cameron said that banks need to lend and with a new approach:

"We need to change the way we measure risk, understand consumers, and make credit decisions. As the recession showed, the systems most banks use today simply aren't ready for the new lending environment."[45]

He said that banks must move away from static measures of consumer risk and customer behavior, and develop adaptive systems that identify changes in behavior more rapidly, estimate likely future changes, and change decisions as a result.

The underlying analytics for measuring risk is predictive analytics. It provides insight on future behavior that can help identify the best action to take on every customer and transaction. For example, in fraud detection, the modeling can detect subtle aberrations from the norm/expected in applications, purchases, claims, transactions, and more. It can detect whether the right person is using a credit card and, if not, indicate to not approve the purchase. Or if the credit limit is reached, it can decide to extend the limit to achieve more purchasing, leading to more lending. It can make loan decisions in seconds through a series of algorithms. It can match personal characteristics with sales campaigns to offer the right product at the right time. And it can do these calculations with reliable precision, embedded in processes in the back office. It can predict whether a customer is likely to switch to a competitor. It creates customer segments based on patterns of similarity and relationship to a specific product or outcome. It can predict customers' future behavior in relation to spending. And much more.

The fuel for predictive engines is data, lots of it, and different kinds of it.

The amount of publicly available data about each and every one of us grows all the time, to the point where an organization with the will, the patience, and the right software can analyze in detail our financial transactions, our habits, our political leanings, our preferences, and our geographic location, among other things.[46]

Perhaps the most instructive view on the new uses of predictive analytics comes from new banking competitors and their disruptive ways to know customers better and steal them away from banks. For example, ZestCash[47] provides loans to people with bad or no credit histories. One of its tag lines is "All data is credit data...and we should be using it all to make better underwriting decisions." It was started by Douglas Merrill, a former chief information officer and head of engineering at Google. It goes beyond a reliance on FICO credit scores, thought to be based on 15 to 20 variables, and looks at thousands of indicators. For example, if a customer calls to say she will miss a payment, most banks ding the customer as a poor risk. But ZestCash has found through its data mining that such customers are actually more likely to repay in full. Another useful signal is the length of time customers spend on ZestCash's Web site before applying for a loan. It is not clear what this might be indicative of but it is a variable in their predictive models.

Wonga[48] is a British start-up that offers loans for very short periods. On its Web site it says that it can "deposit up to £400 in your bank account by 22:31 today," or five hours away. It also looks at a huge range of data such as e-mail address and social-network sites, to make credit decisions practically in real time.

Another firm, Cignifi,[49] mines mobile-phone records. It says that "anyone with a mobile phone can be credit scored." It looks at variables such as the time when calls were made, their frequency, and the whereabouts of the callers for clues about their likelihood to repay loans.

Tesco,[50] a large British retailer, like other retailers described in the retail chapter, uses its customers' purchasing data to send targeted coupons. For example, "when a household starts buying nappies, signaling the arrival of a new baby, Tesco usually sends discount vouchers for beer, knowing that the new father will have less opportunity to go to the pub." The firm has banking ambitions. It offers credit cards and loans and plans to introduce full bank accounts. Given the depth of its databases, it might well assess the creditworthiness of its customers on the basis of their grocery shopping.[51]

CoreLogic is a full-service financial information company that provides information on personal characteristics related to risk assessment, as with credit applications and collection procedures. They maintain 215 million consumer credit records representing more than 35 million unique customers.[52] What's most intriguing is the breadth and depth of the information they collect and synthesize. They collect more than 100 credit variables. They derive information from traditional sources, including credit bureaus, public records, and bank accounts. And they collect it from nontraditional sources, including payday loan companies, rental purchase stores, credit card companies, consumer finance businesses, nonprime auto lenders and credit unions, eviction notices, bankruptcy proceedings, criminal history, mortgage deeds, and many more. These data are all provided to client companies to help them cover risks more accurately.

There is also a push from credit raters in getting data for a "secondary underwriting screen for credit" that includes data on paying consistently over a substantial period (as with rental payments), buying life insurance, education attainment, employee effectiveness, parent effectiveness, and sticking with an exercise program. And the way to get the data is through "meaningful dialogues and listening."[53] Whoa. Meaningful dialogues? What has come over bankers? Isn't the readily available big data enough? How are they going to find out about their customers stick-to-itiveness with exercise programs? Well, they are going to ask them. And they need the data to recruit, retain, please, and optimize their customers. They know that if they do not know their customers well, competitors will steal the space. And they do this despite stringent privacy concerns related to consumer protection. They must ensure that they are using fair and consistent lending practices, sharing information appropriately, and securing personally identifiable data.[54]

Getting back to the HICS. Banks are very clear on what their primary outcomes are: profitable loans. They have honed analytics to determine the capabilities of their customers to follow through on their obligations. And they are interested in extending the period of assessment from a static determination to one that is more dynamic and continuous over time. They even understand the correlates of successful outcomes beyond the traditional confines of financial metrics. In so doing, both the bank and its customers can be partners in the success that both want. And this will require, perhaps, a good deal more sharing of personal information, which might be possible only through a trusting relationship. And this might be the fly in the ointment for banks, but not necessarily for healthcare.

Banking can teach healthcare that it is important to understand and manage the capabilities and obligations of its customers. In healthcare, these obligations and capabilities are hardly taken into account. What if the composite HICS score were listed on the patient's medical record? A low score would indicate to the caregiver that the patient was unlikely to follow through with recommendations on how to get healthier. The score could be decomposed into its component parts such as these:

- Inability to engage in care due to finances such as underinsurance and low income
- Lack of family support
- Previous noncompliance with treatment plans and medication regimens
- Poor education
- Cultural issues
- Lack of transportation
- Lack of availability of nutritious food
- Restrictive health-plan benefits

It is understandable why such a score does not exist in healthcare. Care is fragmented. Achieving health is not always the goal. Payment policies support individual rather than coordinated approaches to care, and focus on how many things were done rather than how well. There is industry fuzziness about defining outcomes and using these in performance measurement of systems. Accountable care organizations can address these concerns. But, for now, let's put the HICS score on the chart and in the values statement for the organization to put a spotlight on the need to attend to people's capabilities and supports, or lack thereof, to improve their health. And then to do something about it.

An example is the organization Health Leads.[55] It strives to ensure that one day all doctors will be able to prescribe solutions that improve health, not just manage disease. These prescriptions include the need for food, housing, health insurance, job training, fuel assistance, and more, which are often more influential in producing health than "traditional" healthcare. Health Leads "connects the dots" between the healthcare system and social systems and achieves improved health results through an innovative business model. For example, they note that "every day in America, doctors prescribe antibiotics to patients who have no food at home or are living in a car." It is not the doctor's job to improve the social determinants of health for society, and they don't have the time or knowledge to address patients' basic resource needs. The HICS can help in this assessment and provide information to caregivers and then coordinate the provision of enablers for better health attainment through organizations like Health Leads.

Conclusions

The unique analytic sweet spots in banking are efficient transaction processing, self-service channels, and risk prediction. The challenges facing the industry are daunting. Analytics can help it lift out of doldrums and to new heights. And the shape that analytics takes in the pressure cooker of reform and business survival should produce analytic resources that other industries can learn from.

The business demands of banking are not unlike those of the healthcare industry, especially health plans. The Affordable Care Act imposed sweeping changes on the health insurance industry on a par with the reforms in the banking industry. Health insurers face a good deal of oversight, changes to basic coverage, and a reorientation toward the "real" customer. Both industries face commodification and an assault from disruptive innovators. The IT challenges and opportunities to support the business are significant. For both industries, a major challenge will be (re)establishing a mutually satisfying relationship with its customers after a long period of relative disregard.

The need to make the business basics work better through analytics is the first order of business in both industries. Both need to respond to a huge list of compliance requirements, and both need to adapt to the changing preferences of their customers in terms of business channels. Mobile applications appear to be the epicenter of transactions and communications.

Both industries seem to be traveling on a similar journey and at about the same point. Banking has a bit of a head start in making the business run better through analytics, but the lessons are not complete at this point.

The real lesson from banking is its depth of experience and techniques to understand risk. The risk has to do with the capabilities and motivation of its customers to carry through on their obligations in the partnership with the bank; that is, to repay a loan commitment. It is acquiring more and varied information to understand its customers better, which will improve loan risk assessment and might improve relationships if it does not breach basic privacy rights. But what is so starkly different in healthcare is the industry's blind spot in understanding people's capabilities and constraints to follow through with treatment "orders" and then to do something about the situation. Rather, healthcare prescribes its solutions to patients and moves on to the next prescription without considering the overwhelming importance of other determinants of health. The one big, unique solution from banking for healthcare is to have an analogue to banking's FICO score, called the Health Improvement Capabilities Score, which should be considered as much as lab values and other findings in guiding the course of care.

6

Sports

One knows much, much more about the performance of a baseball player, who entertains us, than about our physician, who directs our care that can mean the difference between life and death.

Introduction and Industry Challenges

The sports industry is about entertainment and pleasing the fans...and the fans love their sports. For example, there were 75 million attendees at 2,420 Major League Baseball games in 2011.[1] There were more than 15 million viewers for game four of the 2012 World Series.[2] Almost 27 million attended a NCAA Division I men's basketball game in 2010–2011.[3] Additionally, there is more ink in daily newspapers on sports than on business and the arts; local TV closes out its nightly program with a sports segment; there is a seemingly endless variety of programming from specialized sports cable networks; and fans are very active in betting on winners.

In terms of total revenues, the industry is relatively small compared to other industries in this book. The "Big Four" leagues in the U.S., the National Football League (NFL), the National Basketball Association (NBA), the National Hockey League (NHL), and Major League Baseball (MLB), bring in about $24 billion in revenue during a typical year, and the average value of a Major League team is about a half billion dollars.[4] The overall sports industry is touted to have revenues of more than $400 billion when gambling, sporting goods, video games, licensed products, endorsements, stadium naming fees, personal fitness programs, and other sports products are taken into consideration. The U.S. healthcare industry is about seven times its size, at $2.7 trillion. Further, there are about 12,600 professional American athletes[5] compared to about one million doctors.[6]

However, the analytics of the sports industry are much more extensive and detailed in one particular way as compared to other industries, especially healthcare. One knows much, much more about the performance of baseball players, who entertain us, than about our physicians, who direct the care that can mean the difference between sickness and health, life and death.

There is a mesmerizing array of performance data on athletes. In baseball, for example, there are traditional batting and fielding stats, such as batting average (AVG) and runs batted in (RBI), and new Sabermetrics stats, such as on-base and slugging percentage (OBS). The data are sliced and diced in all sorts of ways. For example, batting stats are cut by periods of time, like a "10 game log"; batting averages by career, postseason, all-star, and miscellaneous; by "splits" such as hits versus left- or right-handed pitchers, whether home or away, day or night, grass or turf; and by a specific pitcher. These stats are displayed in various media, from baseball cards, to newspaper box scores, to stadium displays, to elaborate running commentary on TV.

Because of the abundance of data, choosing a baseball player is so much easier than choosing a physician. When you go to a health plan or medical group Web site to choose a doctor, the information provided is comically bereft of relevant detail. One can learn about the medical school attended, specialty, languages spoken, gender, and office address, but absolutely nothing on the performance of doctors...not even how many times the doctor has been at bat, never mind whether the doctor is a champion in his or her field. Even Consumer Reports, which rates almost every product imaginable, does not rate doctors. Commercial Web sites that attempt to rate doctors provide spurious results given small sample sizes and generalizations from generally unhelpful data. Medical societies will let you know whether a doctor has had a malpractice suit. And states concentrate on informing their citizens about the performance of hospitals and health plans, but not doctors.

It's not that there is no information on doctors. There is a plethora of it. It is just not transparent to consumers. I am struck by the contrast: When a baseball batter comes to the plate, all his vital statistics are displayed for all to see. But when I go to my doctor, I feel uncomfortable asking serious questions like how many colonoscopies has the doctor done and what is her average procedure duration. And if I ask, I am sure to be dissatisfied with the evasiveness of the answer. And I certainly do not see a digital sign above the receptionist with this important performance data displayed.

Another unique aspect of the sports industry is the customers' huge appetite for gambling. The FBI estimates that more than $2.5 billion is illegally wagered annually on basketball's March Madness playoffs. Nevada's legal sports wagering represents less than 1% of all sports betting nationwide, at $2.88 billion. The National Gambling Impact Study Commission (NGISC) estimated that illegal bets amount to as much as *$380 billion* annually.[7] It might be that the availability of stats provides the "information infrastructure" required for fans to make rational decisions about betting. Fans might even have their own algorithms for predictive modeling to improve their chances. But it is just as likely that using a data-driven approach to gambling decision making is just as rare as making (mostly intuitive) decisions about one's stock portfolio. Nevertheless, the sports industry feeds on a huge stockpile of data to enhance the entertainment features of sports, and these include gambling.

The next section offers an overview of the key business drivers of the industry, including pleasing the fans, acquiring talent, improving performance, and managing the enterprise. Following that, the major analytic sweet spots are discussed.

Key Business Drivers and Industry Strengths

There are four industry strengths that define sports: please the fans, acquire talent, improve performance, and manage the enterprise.

Please the Fans

In sports, it's all about pleasing, indeed delighting, the fans. Fans buy the tickets, associated merchandise, cable news sports programming, "fantasy" sports video games, and more. And they gamble a lot of money on the games as well. It's all about "play" in sports. "Play ball" and "play-by-play" describe the action in this industry. In healthcare, doctors "practice" medicine. Politicians "campaign." Bankers "invest." So, by the very nature of the action word that describes it, sports is about play and people enjoy the opportunity to do it as often as they can.

The most important way to excite fans is to win. Winning generates more money for the team across the entire product portfolio. Fans love winning teams, especially "come from behind" winners like the Oakland A's, as portrayed in the book *Moneyball* in the early 2000s, and the Detroit Tigers in 2012. Fans love the stars as well. For example, "Linsanity," caused by the Knicks basketball team Taiwanese star Jeremy Lin in 2012, resulted in "every metric going through the roof," including 70% higher merchandise sales (and growing after 50 new Lin-related items were added), record TV ratings for the Garden's 40-year-old MSG cable network, 750% higher Web traffic on the Knicks Web site, and video views up 2000%, all since the team win streak that coincided with Lin's first seven games as a starter.[8] But star power is second fiddle to winning power. A great star on a losing team cannot fill the seats despite his performance and likability.

"Wins" is the solitary outcome measure by which sports teams are measured. Winning is a function of a number of factors. Chief among the success factors are recruiting the right players, keeping them healthy, coaching/learning to improve performance, and managing for team success.* Each of these areas is discussed a bit later in this chapter.

Fans are very involved and even emotionally engaged in the sports, much more than in other industries covered in this book. For example, more Americans (a huge 45%) filled out an NCAA tournament bracket in 2012 than voted in the 2010 midterm elections.[9] As stated earlier, fans spend a lot of their leisure time attending events and viewing them on TV. In contrast, in healthcare, 85% of Americans rate their overall health as good, very good, or excellent. Only 10% have 15 or more days in poor physical health. So, the vast majority have little contact with the healthcare

* Note that the book concentrates mostly on team sports because individual sports like tennis and golf are much less invested in analytics, perhaps because the "enterprise" is one person. Individual athletes such as Tiger Woods in golf and Roger Federer in tennis can achieve higher revenue levels than athletes in teams perhaps because they are unshackled by the limitations of drafts. They can highlight their brand more easily because they are more visible and get all the attention; that is, they play all the time and are not up to bat every other inning and do not get a fly ball only a few times per game.

system except for checkups or visits for relatively minor illness. And most would not want to use any of their time, leisure or otherwise, navigating and engaging with the healthcare system.

Another form of engagement is exemplified by gambling. As cited earlier, total sports gambling is estimated to be between $300 billion and $400 billion a year, which far exceeds all other industry revenue categories. Compare this to the total receipts for club business for the major sport teams at $24 billion. Gambling generates more than ten times more revenue than the actual core business. And customers will go to extremes to do it by seeking out illegal venues.

Data plays a vital role in pleasing the customer in sports. It is an integral part of the vocabulary. The athletes have their performance digitized and laid bare. And more and more of it is used to transform the business from a "club" to a lean-and-mean corporation as described in the "Manage the Enterprise" section later on.

Acquire Talent

The prices paid for high-performing talent in sports are astronomical compared to the other industries covered in this book. For example, according to Forbes, the 100 highest-paid athletes made a combined $2.6 billion in the year ending June 2012, or an average of $26 million each, for salaries, bonuses, endorsements, and so on. Examples of the highest-paid team-sport athlete's salaries (alone) paid by sport clubs included Alex Rodriguez ($32 million, baseball), Larry Fitzgerald ($35 million, football), Cristiano Ronaldo ($20 million, soccer), and Kobe Bryant ($20 million, basketball).[10] Endorsements for the highest-paid athletes can double the amount made from salaries. For example, the top-10 highest-paid athletes made an average of $25 million in salary and bonus and an additional $28 million in endorsements, for a total of more than $50 million a year.[11]

The top-ten corporate CEOs probably made this much. And there is a good deal of regulatory and stockholder demand for transparency about total compensation. Politicians certainly do not make this kind of money from their day job, although it seems that only multimillionaires have enough money earned from other careers to invest in getting a low-paid public job. Doctors and other healthcare professionals do not make this kind of money, although some CEOs of provider and health insurer companies might do so. So nowhere else is the "winner take all" American story more apparent than in sports.

But the teams that pay the highest salaries and have the best stars are not always the teams that win the most. And there are well-known examples of baseball clubs that have low budgets for talent and have gone on to the playoffs. All businesses increase profits by getting the most revenue at the lowest costs, and talent costs are often the most substantial investment a business makes. As teams are increasingly managed by business executives, they focus on exploiting market inefficiencies by getting the best talent at the lowest price. And to do that they need the right numbers.

For example, the book *Moneyball* details how the restricted talent budget of the Oakland A's led to innovation in using analytics to predict performance and to acquire high performers at the lowest possible expense. For example, the team was able to draft perhaps the best hitter in country with the 218th pick of the baseball draft for a very low price.[12] And in the years following the Billy Beane data-driven approach to acquiring talent, the team went from winning 87 games in 1999 (before) to 91 games in 2000 and 102 games in 2001, and made the playoffs both years.[13] Many baseball teams followed in the footsteps of the A's to bring in analytic talent to help find potential high performers whom no other team had discovered, for whom they could pay a relatively low cost of acquisition. To further erode the importance of traditional star power and its demand for high salaries, analytics is developing a new understanding of the ingredients of what makes for team wins and decomposing these into component parts and buying the pieces more cheaply than the whole stars.

Improve Performance

Although most professional athletes have trainers and coaches, there has been decidedly less effort devoted to improving performance relative to picking the right talent. Players certainly "review the tapes" of their own performance and those of players on the opposing team to get tips on how to improve play and react to other players. And there is more attention now on the composition of players and how they interact for optimum team performance. But the prevailing sentiment has been that it's all about the player's responsibility for performance, and when that does not meet expectations, the player is traded or fired. Billy Beane said, "You don't change guys; they are who they are...you build the car in the beginning of the year and after that all you do is push it down the hill."[14]

This is changing partly because of the availability of new and innovative performance metrics and because the management of teams has become more business oriented and less influenced by the jocks and somewhat more by the "statheads." In the section "New Data: Nature of the Sport," we will see that every frame of play action is videoed, recorded, plotted, and modeled. The only thing missing is an earpiece in athletes' ears to tell them what the next move should be. So the technology and the analytics are ahead of the headlights of management in terms of how to use these data to improve play in real time. But, as with other industries, the organization that does the right modeling and can predict and react to play action will have a decided advantage. What is clear, however, is that in this industry, the vast amounts of new data and its possibilities have not quite been assimilated into decisions about behavior change on the court or in the practice rooms. But that is coming to the industry.

One thing is for sure. All industries are in dynamic markets and nothing stays the same. Sports is all about the talent. Finding the best talent. Keeping them healthy. And it should also include bringing out the best of each athlete's unique talents by understanding how to improve the individual's and the team's performance.

Manage the Enterprise

Indeed, winning is what fans want and it is the primary driver for generating revenues. In addition to acquiring the best talent and improving performance to produce wins, the front office needs to manage the enterprise to make a profit in a variety of ways, such as TV rights, product merchandising, and premium seats. Just as with performance improvement, there are new opportunities driven by the data for improving the enterprise. For example, fans seem to have a huge appetite for engaging in and spending money on sports such that a concentration on customer analytics, referred to in sports as "fanalytics," might lead to new revenue opportunities. A few examples:

- Understanding fan behavior (customer analytics in other industries) to drive sales, pricing, and loyalty.[15]
- Fine-tuning data-driven player evaluations into player valuations and systematically determining a fair price for talent.
- Moving from a preoccupation on individual performance to team performance and overall club performance.
- Building analytics into the physical structure of the game. For example, if a new baseball stadium had a shorter distance to the fence, would it benefit from the hitting behavior of its players?

The management culture of sports is also changing toward a more—dare I say it—scientific approach to management. Jocks have run the "clubs" for a long time. They relied on their sports intuition to guide decisions. Experience counts, absolutely. Baseball recruitment has long been dominated by scouts who use their experience to judge whether a player has a "good face," which is shorthand for whether he catches and swings like the guys in the big leagues. But Boston fans who believe that their Red Sox club had suffered a losing streak under the Curse of the Bambino for decades wanted a change. The club hired Bill James, the godfather of baseball stats, to provide a new data-driven perspective on talent. (Note, however, that the club suffered one of its worst records in history in the 2012 season.) The nerds are still regarded as extraterrestrials from a very strange planet, but a good deal of experimentation is underway in using data to address the "proof of concept" of data-driven decision making.

Analytics Sweet Spots

The three sports analytics sweet spots include performance measurement and transparency: players and teams; predictions: wins and best plays; and new data: nature of the sport (see Figure 6.1).

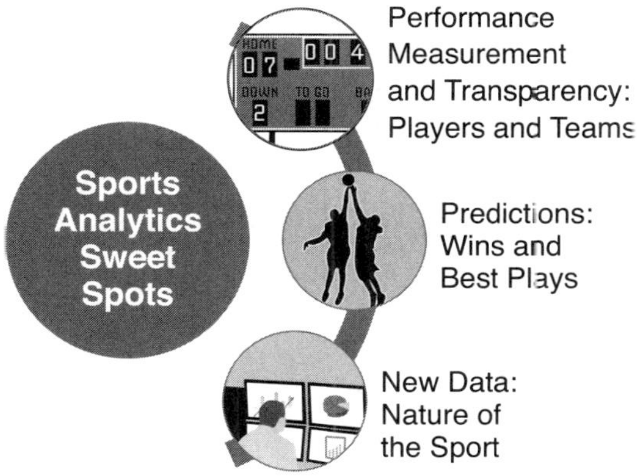

Figure 6.1 Sports analytics sweet spots.

Performance Measurement and Transparency: Players and Teams

Individual performance metrics is the unique analytics sweet spot of the sports industry.

When I was a kid, I knew all the stats of my favorite baseball players through baseball cards. There were five cards to a pack along with a stick of baseball gum. I bought so many that an alternative use was to build huge houses out of the cards. I also remember my son's fascination with buying and selling cards. One could go to *Toppers* magazine and find the current value of a card. There were many baseball-card shops where one could buy and sell cards. The card market kept pushing prices higher and higher and then fell apart, perhaps because all the information could be obtained over the Internet.

I have a card for Darryl Strawberry from 1985 when he played for the Mets. The card displays performance metrics on his statistical "dashboard." It includes his physical, demographic, and draft status, as well as a wide array of his batting statistics, including at bats (AB), runs (R), hits (H) and types of hits, runs batted in (RBIs), and batting average (AVG), by club and totaled for his entire career. All the dashboard metrics are on one piece of paper about the size of a small postcard.

The basic stats are the same as they were more than a hundred years ago, when the first newspaper box score included stats for each player, including runs, hits, putouts, assists, and errors.[16]

In addition to the newspaper and cards, the stats were flashed on the stadium scoreboard as each player got up to bat, went out to pitch, or fielded a play. And, of course, the running action on TV included qualitative data from the sports commentators and quantitative statistics that were referred to during the play action.

Today, ESPN and other outlets publish voluminous stats on player performance for major sports including the NFL, MLB, NHL, NBA, soccer, and more. In addition to the baseball-card dashboard of key statistics, it slices and dices the performance in myriad ways, as discussed earlier.

Except for the extensive slicing and dicing, the performance measures are pretty much the same as they were decades ago. However, there are some interesting additions, including on-base percentage (OBP), slugging percentage (SLG), and on-base and slugging percentage (OPS). These three metrics are from a new school of baseball metrics called Sabermetrics.

Sabermetrics is an acronym for the Society for American Baseball Research. These stats focus more on the team and how a player contributes to winning. Sabermetrics argues that runs win ballgames and that a player's worth is directly proportional to his contribution in enabling his team to score more runs than opposing teams. For example, through its relatively rigorous analysis, it demonstrated that OBP and SP are better indicators of team success than traditional measures such as batting average and runs batted in. Even getting on base with walks is important because if a player is not on base, he cannot produce a run. Another scientific approach to measurement is VORP, or value over replacement player. This is the number of excess runs a player is worth over a replacement player at the same position. It is analogous to education metrics of determining expected student outcomes based on socioeconomic and other factors and then attributing the excess performance to the skill of the teacher. And in healthcare, it is analogous to risk-adjusting outcomes based on the severity and other characteristics of patients and linking this to pay-for-performance awards to physicians.

The seismic shift that occurred with Sabermetrics and data-driven approaches to baseball exemplified by Billy Beane in *Moneyball* is the change in the perspective on baseball from a hobby/pastime/entertainment to a business. Business success derives from winning (and producing revenues) and keeping costs low. A business intelligence point of view demanded that the factors associated with winning be measured, monitored, and acted on. For example, the general manager of the A's used the data to (micro)manage the business. "I got my reports...I can see they aren't taking any walks. I called the manager and said, they go up or you're fired. And they went up."[17] The real team business victories arose when the team achieved great outcomes, for example, a series win, at the lowest costs, using relatively cheap draft choices, to optimize profits.

There is a new direction in sports performance measurement that directly arises from the capability to measure important aspects of sports through new technologies. (This technology is discussed more fully in the section "New Data: Nature of the Sport.") It is possible to capture every event on the court or playing field through video capture by minicams and then digitize it into a spatial grid, including the "latitude and longitude" and the height. These data can be annotated to include which player made the shot or fielded the play or defended the play. It creates a huge amount of data, which is the equivalent of big data in sports and tests the capability to use unstructured data for performance measurement and performance improvement.

Let's look at basketball. As in baseball, basketball has a standard dashboard of performance metrics. The analogue to batting is shooting, and the stats include points per game, field goals made, field goals attempted, and field goal percentage (FG%). FG% is a summary measure of the shots made divided by the shots attempted.

The top-ranked FG% players (all centers) for 2011–2012[18] were as shown here:

Tyson Chandler	NY	.679
Dwight Howard	ORL	.573
Nikola Pekovic	MIN	.654
Andrew Bynum	LAL	.558
JaVale McGee	DEN/WASH	.556

But is this simple metric the best or most complete indicator of shooting capability? Kirk Goldsberry, a sports and healthcare data scientist who recently graduated from MIT, wanted to address whether a shot was good or bad and whether players who shot and made baskets from many areas on the court were more valuable for team wins. His reasoning/hypothesis was that it was true because diverse shooting places on the court would be less predictable and less defended.

Goldsberry's first approach was to develop new performance metrics called "spread" and "range."[19] Using data from every NBA game played between 2006 and 2011 that recorded every field goal attempt from any of the 1,284 possible shooting cells of the court, he calculated spread percentage as the percentage of the scoring area in which a player has attempted at least one field goal. The more important metric is range percentage, which is the percentage of the scoring area in which a player averages more than one point per attempt.

His ranking of range percentage for NBA players during this period is shown here:

1.	Steve Nash	31.6%
2.	Ray Allen	30.1%
3.	Kobe Bryant	29.8%
4.	Dirk Nowitzki	29.0%
5.	Rashard Lewis	27.6%

Goldsberry also uses visualization graphics to display range percentage. The graphics reveal shot attempts and points per attempt. Larger squares indicate areas where many field goals were attempted; smaller squares indicate fewer attempts. He also uses color to indicate the average points per attempt for each location.

For a performance metric to be useful, it needs to show an association with an outcome as OBP is with wins in baseball. In healthcare, this is a prime criterion for endorsing performance measures by the National Quality Forum. For example, improvements in HbA1c are important for the clinical outcomes of people with diabetes. It is unclear whether these technology-induced sports measures are associated with outcomes. They need more testing. The data might be useful for the defense to know where a player is likely to shoot from. However, it would seem that the elaborate video, spatial, and specialized analytics are not totally necessary for this purpose. If all players had earpieces, it might be useful to guide defensive play in real time.

So, individual performance measurement is the unique analytics sweet spot of the sports industry. It has been around since the inception of the industry. Clearly, the fans love their sports and the stats help them follow their sports heroes and teams. It is mostly about entertainment for the fans. It is somewhat unfathomable to me how much attention is paid to describing every action of sports through so many channels by so many pundits with so many stats. Maybe the stats were necessary to fill the void in the action on the field such that the commentators had some "fill" for programming.

Other industries have publicly available statistics that consumers use on a regular basis but with a difference. The most obvious is in finance, which publishes stock price information in daily newspapers and displays and discusses it 24/7 on specialized cable news shows. Many people are actively engaged in managing their own stock portfolio since the wholesale move to 401k plans rather than defined benefit plans in which professionals made decisions for them to maximize returns for their pension. The box score measures are relatively simple, including high, low, and close prices and year-to-date return on investment. Underneath the box score, however, is a mesmerizing array of performance information. But the difference is that the stock stats are at the organization (corporate) level and not the individual broker level as in sports.

There are performance metrics determined by polling in politics on the standing of candidates. The statistics cover a range from a general indicator, such as the percentage likely to vote for a candidate, to somewhat detailed opinions on favorability issues, for example, "Who do you trust?" These are different in that the stats are based on opinion rather than "fact"—for example, strikeouts—and on sampling rather than all the actual data. Voters are interested in the race standings and they do have political heroes that they follow. But the statistics are relatively few in number in politics, and the degree of engagement with the political performers is more remote and even involves angst rather than the fun factor in sports.

In healthcare, there is nothing approaching this fascination with the behavior of its individual *or* organizational counterparts. In fact, people are more *un*engaged in healthcare than they are in sports or politics in terms of information seeking. Those who are sick and smart might use the Internet to help them understand their disease and evaluate possible treatment options. But, mostly, people do not have access to price information or quality information and they cannot shop for care either because their choices are limited or predetermined or because they have so little skin in the game such that it hardly seems worthwhile. A few state governments do provide information on the performance of health plans and hospital providers, but healthcare consumers do not use it for various reasons, not the least of which is that it is hard for the consumer to understand how the information can be of benefit to them.

What can baseball and other sports teach healthcare in regard to performance measurement? And why do we know so much about our sports heroes and so little about our doctors? I mean, after all, sports is just entertainment. Healthcare is about life and death.

As we have discussed, there is a huge effort to recruit the right athletes, and it is partly data driven based on past performance but mostly influenced by the intuition of scouts. The stakes are high in financial terms for selecting a great athlete. The acquisition and ongoing cost is high—in the millions for top-tier talent. And teams that are in the running for a series win value an extra game win or two in the millions of dollars.[20] Therefore, from a financial calculus there are high table stakes just to be in the game, that is, to operate a major league club, and these get higher if the team wants to be a contender for a series win where the rewards are much higher.

Compare the data available for the selection of a baseball player with that available for the selection of a doctor. Consumer Reports rates everything from toasters to cars, but it does not rate doctors. It states that "choosing a doctor is one of the most important things you will do in your life. You will trust this person for his or her expertise, advice and to administer treatment that may save your life."[21] It does not provide any ratings but advises you on how to go about selecting a doctor, including getting a referral, knowing your medical issues, asking how many procedures of the type you need he or she has performed, and checking how well the managed care plans that the doctor is involved with does with rating groups like the National Committee for Quality Assurance. On this latter point, it would be like getting the win-loss stats of the college team of the prospective player.

So if the über-rating group does not rate doctors, who does? The health insurance plan and medical group are where one chooses a doctor. What data do they provide? My doctor is with Harvard Vanguard Medical Associates. The data about him on the webpage includes when he joined the medical group, where he went to undergraduate and medical schools, where he did his internship and residency, whether he is board certified, what hospital he is affiliated with, and whether he has a faculty appointment. Again, comparing to baseball recruitment, is it helpful to know what college the player attended, where he played after completing school, whether he has received awards, and whether he gives batting clinics?

How about the government? After all, doctors should be monitored relative to the public's safety. The information from the Board of Registration in Medicine[22] and Mass.gov[23] provides the same information on my doctor and a bit more. It provides his license number and renewal date, what insurances he accepts, whether translation services are available, his business address and phone, his professional publications, and malpractice information. But, I still do not know whether he knows how to treat my skin cancer.

Finally, I went to Healthgrades, which rates doctors, and I found that "100% of Dr. X's patients would recommend him to friends and family." That sounds good until I read the small print and see that this is based on two surveys. I also see that he is on the Healthgrades Honor Roll, which includes doctors "who are board certified in the specialty they practice, have never had their license restricted or revoked, have no malpractice claims, and are free of state or federal disciplinary actions." Okay, so I am happy that he is not being sued or disciplined by the Feds and that he is actually an oncologist and not a gynecologist. But does that get him on the "honor role"? Seems like a pretty low bar to achieve noteworthy status.

To make matters a bit more convoluted, the data on individual doctor performance *is* collected and is known by the medical group and insurer, but it is just not made available to the public. So in baseball there is full transparency about individual performers but in healthcare there is none. Why? There are valid concerns that sample sizes are not sufficient to adequately measure individual doctors. But it doesn't seem too far-fetched to count "career at bat attempts at CABG surgery" for doctors who perform it. Even readmissions for certain conditions are only reported at the hospital level through the CMS Hospital Compare site. And there are concerns that patient ratings are subjective and not valid. It is clear that the research methodological perfectionists have conspired with the medical power establishment to protect doctors from data exposure. It is also clear that there is no compelling consumer demand for doctor ratings.

What is it about this reluctance to hold teachers and doctors accountable for outcomes? Education is somewhat ahead of healthcare in this respect; state laws and the Department of Education's "Race to the Top" initiative mandate that at least a portion of teacher evaluations be derived from "risk adjusted" outcomes based on students' performance on standardized tests. In healthcare, pay-for-performance programs reward doctors when specific treatment guidelines are followed, such as performing appropriate tests for the care of people with diabetes. And accountable care organizations funded through global payments will need to have metrics on overall outcomes related to keeping people healthy, treating them appropriately to maintain health and curtail disability, and reducing costs in the process in order to be competitive. However, these metrics are made transparent at the organizational level like the hospital or health plan, or perhaps at the medical group level, but not at the individual level. The organizations that do business with the physician, the provider organizations and health plans, will have information on the clinical and business outcomes of physicians and can use that information to recruit, retain, and improve the performance of individuals.

Maybe the real take-home message is that the overwhelming need in healthcare, as in sports, is to concentrate on outcomes, to have more coordination among many different players, and to actually reduce the concentration on the individual doctor, especially as it regards payment. Fee-for-service payment actually perpetuates individualized action rather than team action. What is really needed is analogous performance metrics of OBP SP and VORP in healthcare whether they are publicly reported or not. After all, healthcare should be a team sport.

Predictions: Wins and Best Plays

Whereas performance measurement is about measuring and monitoring what has happened in the past, predictive modeling is about the future and can be of tremendously more business value. For example, what if teams could pick talent based on the likelihood of solid performance in the future, if pitchers knew what the right strikeout pitch is for a batter, if shortstops knew where to position themselves for each hitter, or if basketball forwards knew where to catch a rebound for each shooter? These questions can all be answered in sports because one of the advantages of the sports industry is its very rich data set, including all types of data collected since the start of the game. In baseball, that would be 140 years of data.

Because of the large sample size, a large number of hypotheses can be generated and tested and statistical significance determined.

The first generation of quantitatively data-driven predictive analytics was Sabermetrics, developed more than 30 years ago for baseball. Prior to that, there was "qualitatively" based predictive analytics as practiced by scouts based on their deep intuition about the nature of baseball players and the "good face." Sabermetrics focused on the outcome (dependent variable) and the variables on the right side of the equation (independent variables) that caused the outcome to get better. Sabermetrics models pick runs as the dependent variable because runs produce wins. A model for creating runs from *Moneyball*[24] is as follows:

Runs Created = (Hits + Walks) × Total Bases / (At Bats + Walks)

This formula emphasizes the importance of whatever it takes to get on base, including walks. The formula gets jiggered and improved by analysts. Weights are assigned to the variables and then a performance metric is developed, such as VORP. The existing basic stats of players are then converted into the new metrics and these values become the determinants for making decisions about player recruitment.

Version 2.0 of predictive sports analytics occurred 25 years later with more sophisticated models that predicted player performance in the next season. The statistics were, and are, published by *Baseball Prospectus*, and sold to the public. It added new dimensions to the equation. One of these is the "aging curve" that determines precisely that a baseball player's peak performance is at age 27. New statistical approaches called PECOTA (Pitcher Empirical Comparison and Optimization Test Algorithm) were adopted, including player similarity analysis, or nearest neighbor analysis, which basically means grouping players according to their stats and plotting likely performance trajectories of the group.[25] It also developed more performance measures such as WARP (Wins Above Replacement Player). Although PECOTA matched or bettered the competing forecasting systems, it did not beat the (competing publication) *Baseball America* predictions, which are based on both scout and empirical data. In fact, the extra wins the scouts identified were worth over $300 million,[26] giving rise to the notion that a "both/and" perspective is better than an "either/or" one when considering the value of scouts versus "statheads" to make better predictions. Additionally, there is greater appreciation that performance metrics are necessary but not sufficient variables and that player characteristics such as work ethic, focus, self-confidence, stress management, and learning ability are important personal characteristics that drive performance.

Version 3.0 goes beyond data for recruitment or predicting player performance to improving the play of the game. The MIT Annual Sports Analytics Conference is the place to be for those interested in sports analytics. It solicits papers on analytics for its annual conferences, and some of the best insights about the field come from its own grads and faculty. Sports teams are mum about divulging anything specific about how they use analytics. Although they appear on panels, they speak in annoying generalities. You have to listen to the academics studying the field.

One paper by Robert Ayer, a 2011 graduate of MIT, addresses what *the best player combination* in basketball is to produce the most wins.[27] It goes beyond the traditional obsession with individual talent to addresses team talent. Ayer's model for NBA success is simple:

Wins = Talent + Coaching + Team Fit

He quantifies each domain of the model. For example, individual talent is calculated according to the following formula (summed across all players on a team):

Talent = ((Points + Rebounds + Assists + Steals + Blocks) – ((Field Goals Att. – Field Goals Made) + (Free Throws Att. – Free Throws Made) + Turnovers))

Coaching is a binary variable with a 1 assigned to the coach who coached the majority of the team's games and a 0 to others.

Team fit is the independent variable in the regression equation of most interest to Ayer. It assesses the most productive mix of talent. He developed 18 clusters of two- and three-player combinations based on player similarity on the following variables: points per game, offensive rebounds per game, defensive rebounds per game, assists per game, steals per game, blocks per game, turnovers per game, personal fouls per game, field goal attempts per game, field goals made per game, free throw attempts per game, free throws made per game, three-pointers attempted per game, and three-pointers made per game. Examples of cluster characterizations and representative players include these:

> Cluster 1: Limited, role-playing centers: Erick Dampier, Tree Rollins
>
> Cluster 2: High-scoring, dynamic guards (mostly two guards, but some threes like Grant Hill), typically not great three-point shooters, or if they are, don't shoot very many: Kobe Bryant, Dwyane Wade, Tracy McGrady, Adrian Dantley.
>
> Cluster 3: Somewhat limited, role-playing backcourt players: John Paxson, Jose Barea
>
> Cluster 4: Wing three-point shooters: Dan Majerle, Shane Battier
>
> Cluster 5: Dynamic, well-rounded power forwards, strong rebounding, dynamic threes: Chris Webber, Pau Gasol, Kevin McHale

His regression results indicated that the best *two-player* combination is a three-point-shooting wing and a high-scoring, high-rebounding center. That delivered 7.59 incremental wins per season. Conversely, having two high-scoring wings reduced wins by 4.05 games per season. For *three-player* combinations, he found that a high-scoring, high-assist point guard, a versatile, three-point-shooting wing, and a high-scoring, high-rebounding center yielded 13.6 incremental wins per season. Conversely, the combination of two dynamic power forwards and a low-scoring, pass-first point guard diminished wins by 8.47 games per season.

This work utilizes a standard multiple regression approach for predicting wins. It holds other variables constant to determine the impact of a given variable. The equation could be refined with additional independent variables, but the results are quite compelling about why and how to recruit for team talent in basketball.

Another example of predictive modeling is a paper by Gartheeban Ganeshapillai and John Guttag that addresses whether data can lead to a batter predicting the next pitch and planning accordingly. Specifically, they developed a model to predict whether the next pitch would be a fastball.[28] Batters certainly look for clues in the delivery and setup of a pitch, and pitchers try their best to hide this information from batters. The batter must make a decision in milliseconds on how to approach the pitch. Additional, real-time insight on the nature of the pitch could dramatically increase batting performance.

Ganeshapillai and Guttag used a machine-learning-based predictor model. The predictor incorporates information about the current at-bat and game situation (for example, the previous pitch, the count, the current score differential, and the current base runners) and some information about the pitcher's tendency to throw a pitch with that property in various situations. They describe their method this way: "We use a linear support vector machine with soft-margin to build a separate predictor for each pitcher, and use the weights of the linear classifier to interpret the importance of each feature."

Their results indicate that for predicting whether the next pitch would be a fastball, the model's classifier predicts 70% of the pitches accurately, which is about 18% better than the status quo.

It is unclear whether an 18% improvement is enough to be useful given a few limitations. For example, what about predicting other types of pitches, such as curveballs? And where the curveball would go? And if it were a fastball, would it be up the middle and high or low and outside? But the real issue is how the prediction would be used in real time. Would the batter need an earpiece? Would the team's data scientist be running the numbers for each batter prior to each pitch and then signaling the third base coach to then signal the batter?

These three versions of predictive analytics in sports demonstrate the progression from the use of the results from recruiting, to predicting athlete performance, to improving the game. It relies on a large database of basic data but does not include potentially useful explanatory variables from other data sets, such as player aptitude and health resilience. As in the next section, the results are not yet useful in real time. Finally, there is insight that maybe going totally quantitative misses the point that it is not the amount of the data that counts but the importance of the data in achieving good results.

New Data: Nature of the Sport

New technologies are enabling the creation of new sports data that could influence team and business performance in exciting ways...in the future. At the moment, however, the major contribution of the new data revolution is to learn more about the nature of the sport rather than how to use it for competitive advantage. There is a lot of hype from vendors about possibilities but little in the form of proof of concept. But the opportunities are important. Some examples follow.

Formula One auto racing is a very popular worldwide sport that claims to have an audience of 165 million people per race. There are 12 teams, playing in 20 races per year, with development budgets of leading teams reaching into the hundreds of millions of dollars. This is a very fast sport and decisions need to be made in milliseconds that can have a significant impact on the outcome of the race.

The analytics idea is to have an expanded "dashboard" that uses "telematics" to collect data on a variety of new measurements of the car to improve performance in real time such as GPS on car position, video views of position and competitive behavior, braking behavior, throttle, and lap times. The availability of all this information *should* improve racing performance. In the future, putting all these data "in motion" such that they inform driver behavior in real time is the hope.

One of the biggest challenges for racing is the pit stop and how to predict it and to extend the time between them. So one concrete example of this new dashboard is understanding tire choice and pit stop behavior that might help in predicting how fast the tires wear and what the pit strategy should be. Transmitting the data from the race car dashboard to the analysts is easy enough. But analyzing it in real time and giving driving directions back to the driver might be a challenge. Yes, race car drivers, unlike most other sports figures, can wear earpieces, so this would help put the analytics into action. Vendors make the usual marketing claims about the importance of "predicting the future" and developing learning organizations, but it appears that the biggest challenge at the moment is building capacity to collect data and to demonstrate its utility.[29]

An example from basketball focuses on the importance of rebounds. The winning paper for the MIT Sports Analytics Conference 2012 was by Rajiv Maheswaran, Yu-Han Chang, Aaron Henehan, and Samantha Danesis. They investigated what happens from the time a missed shot is released to the time it ends up in a player's hand.[30] The importance of rebounds to team wins is that rebounding is regarded as a key success factor, especially in championship play. Specifically, if the offense can capture the rebound and have another shot attempt, that increases the chances for more points and the win. There is a performance metric called the credited assist that documents a rebound ending up in a player's hands. The paper fundamentally seeks to discover the *nature of rebounds*, to describe where they go and where the best place on the court is to get them. It can lead to refined performance metrics and provide key variables for predictive modeling.

The enabling technology is the STATS Inc.'s SportVU optical tracking system. It records the coordinates of all players on the court at 25 frames per second. These data can be supplemented with relevant play-by-play events. Additionally, the ball can be tracked for x, y, and z coordinates. The z coordinate is the height of the ball. The authors also build models on where the ball could have been rebounded as opposed to where it was rebounded.

The authors conclude that the best locations to attempt a field goal to maximize the chance of an offensive rebound is close to the basket or the three-point line rather than midrange shots. This is due to the trajectory of the ball and typical positioning of offensive players in these scenarios.

So this helps the field understand the nature of rebounds but it is hard to see how this can improve play. Should all offensive shots be from the three-point line or under the basket to maximize rebounds? Are the rebounds followed by successful shots? How might the data instruct play in real time? The obvious point from the work is the ability to track everything that occurs on the court with the precision of physics.

As in basketball, there are camera systems that track the infinite details of pitching and fielding in baseball. PITCHf/x is a pitch tracking system, created by Sportvision, installed in every MLB stadium. This system tracks the velocity, movement, release point, spin, and pitch location for every pitch thrown in baseball, allowing pitches and pitchers to be analyzed and compared at a detailed level. Two mounted cameras in each stadium are used to track each pitch and establish the aspect of the pitch. Similarly, Fieldf/x is a motion-capture system created by Sportvision. It uses four cameras hung above the field to track players and the ball and log their movements.

An example of the utility of PITCHf/x is in comparing how umpires' calls fare with the "factual" strike zone as determined by PITCHf/x.[31] The data show the tendency of umpires to call actual strikes as balls in the upper-left quadrant of the strike zone.

This type of quantitative data should be helpful for coaching pitchers. Pitching technique could be deconstructed relative to the results achieved, and then mechanical adjustments could be made to achieve different results. It's unclear how this would actually impact pitching improvement more than a straight video, but it certainly creates lots more data points.

With Fieldf/x it would be possible to know about the positioning of fielders and reactions, throws, base-running actions, jumps on hit fly balls, how hard the ball was hit, and how long the ball was in the air.

The technology is a great improvement over how these data were estimated previously. Previously, video techs would watch videos of every game a number of times and manually enter data on each batted ball, including ball direction, distance, and type (grounder, fly ball, line drive). The new Fieldf/x way automates the previous data collection and adds more data points with each game reported to produce 2 terabytes of data. Talk about big data in sports!

It is not clear at this time whether analytics put to the data produce more wins and provide a competitive advantage. One general manager said, "We don't get paid to advance the state of the art in analysis; we get paid to put a winning team on the field."[32] There is the belief that getting a jump on the new fielding data could be worth an extra win or two per season for a major-league team.

A final example focuses on injuries, their impact on wins, and how they can be prevented in rugby.[33] Rugby is a brutal game by any definition and injuries are quite common (as are missing teeth). The Leicester Tigers rugby club in the United Kingdom gathers data on players' performance and fitness, as well as their vulnerability to injury. They approach the measurement of vulnerability in two ways:

1. They record every event involving player collisions, leaps, kicks, and sprints.
2. They measure the players' activity intensity through a small monitoring device worn during games.

Using these data on vulnerability, they hope to discover each player's fatigue threshold. That way they would know when to pull a player from the game, when this indicator of vulnerability to injury hits the danger mark.

The new data being generated for the sports industry does not have a specific pathway to improve business. It is rather like basic research in understanding the nature of the basic elements of sports. It is generated by technological capability. And a lot of experimentation is underway that hopefully leads to proof of concept that can then take the data from espoused theoretical uses to valued practical applications.

Conclusions

The unique analytic sweet spot in sports is performance measurement at the individual athlete level. This has been a feature of the industry from the beginning and is largely to satisfy the appetite of fans. The book makes the observation repeatedly on the paradox that consumers know much more about their sports heroes than about their doctors.

The difference between the two industries in this regard is the customer demand for the information. Fans want it. Patients do not. Fans are actively engaged in sports. Patients are not anywhere near as engaged in healthcare. For example, few patients are engaged with shared decision making about elective procedures even though this decision is more personal than medical. Part of the lack of engagement is the overmedicalization of healthcare and the assigned role of the patient as a dependent observer rather than an active participant.

The important decisions about healthcare are not made at the insurance level or even the hospital level, for which there is abundant performance data. In fact, consumers have little say in these matters. Although these data are publicly reported, consumers do not use them. People relate to their doctors and not to their insurance companies or hospitals. Again, there is a paradox in healthcare that the information that consumers really need is withheld and the information that is available to them is not of much use.

Doctors account for 80% of spending and a commensurate influence on the decisions made about the care provided. And the data on physicians is abundant, especially with electronic medical records. Most of these data that could be used for performance-measurement purposes and other important purposes as well are currently treated mostly as "digital waste," a byproduct of a specific purpose; for example, the clinical interaction.

There is certainly documented variability in physician performance, although most consumers regard physicians as providing equal quality of care. Patients are mostly loyal to their own physicians. And many feel that it is inappropriate to grade them. Much of medical care is routine with a very predictable clinical course and clinical outcomes—as for headaches, upper respiratory conditions, high blood pressure, and the like—and perhaps are not worth measuring relative to performance. But many patients have very complicated conditions, can be injured, do not receive the care they should, and die because they do not have the option or information to choose their physician.

One possible way that performance measurement of physicians will change is through customer-controlled information sharing, as with social media and sites like TripAdvisor for healthcare. With these methods, routine evaluation and communication of that information happen naturally, rather than through sporadic customer satisfaction surveys, as with CAHPS or Press Ganey. If people are invited to be actively engaged as a part of the care process, as well as the social communication process, they might well define what information is collected and how it is used.

On the flip side, healthcare might have it right and can teach sports something to improve its industry. Healthcare, like sports, should be considered a team sport, and measurement and management should be directed accordingly. Putting the metrics on the team's performance rather than individual actors might lead to better coordination, efficiency, and outcomes in healthcare...and in sports. And sports team management is beginning to realize that as well. Although sports generates a lot of data and it appears to know a lot about data on the surface and in TV commentary, its use of advanced analytics to improve the competitiveness of the teams and to produce value for the business is in its infancy.

7

The Top Healthcare Analytics Adaptations

There is a common thread running through the analytics adaptations. Simply said, the way to improve outcomes is to ramp up the involvement of people in the co-production of their health.

This chapter presents a distillation of the ideas generated throughout the book into seven healthcare analytics adaptations. These adaptations emerged from the adaptations discovery process described in the first chapter. The ideas chosen for development into adaptations are the truest representation of the industries' sweet spots and the most amenable to shaping into solutions. They also happen to address significant challenges in health and healthcare.

Seven Adaptations

The seven analytics adaptations cluster into three important areas of health and healthcare, including population health, patient engagement, and provider accountability. Figure 7.1 shows how each of the adaptations maps to the clusters. The adaptations for Population Health include obesity detection, well-being, and my dashboard. For Patient Engagement, the adaptations include radical personalization and capabilities and support. Provider Performance includes the adaptations of peer-to-peer provider performance ratings and team-centered outcomes.

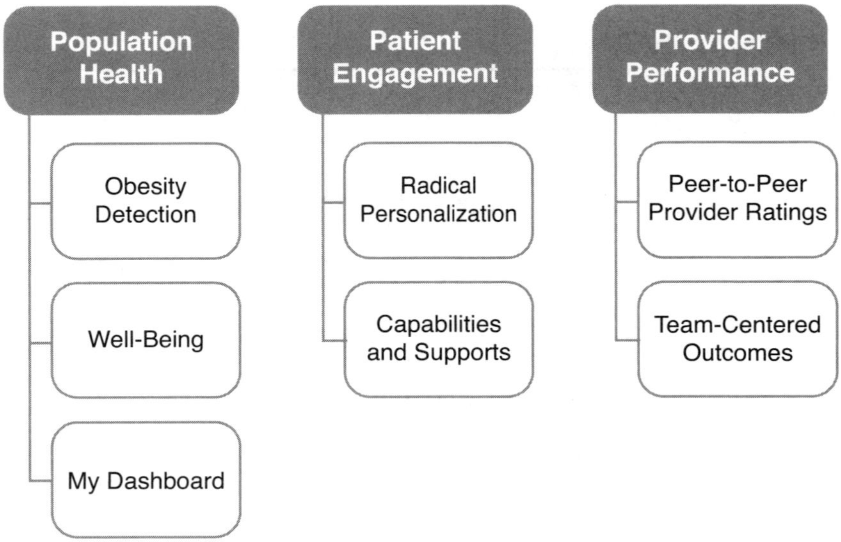

Figure 7.1 Seven healthcare analytics adaptations.

Each adaptation is described in terms of the business need and business goal, the industry derivation including the relevant analytic sweet spots, the analytics goal and associated methods, and the major challenges for adopting the analytics into practice. Note that the next chapter presents a how-to framework for evaluating and progressing an adaptation through the adoption decision stage of the Innovation Pathway, a framework for the journey of an innovation from the idea stage to its successful implementation and reinvention. A hypothetical case study addresses the many issues at play for successful adoption and uses the Innovation Adoption Factors Model as a guide.

Obesity Detection

The first analytics adaptation addresses the challenge of reducing the burden of chronic diseases through early identification and intervention (see Table 7.1).

Table 7.1 Analytics Adaptation #1

Population Health	Obesity Detection
Business need:	Reduce the burden of diabetes in terms of lives and dollars saved.
Business goal:	Reduce incidence of diabetes through early identification of obesity followed by provision of weight-reduction intervention programs.
Industry sweet spots:	Retail: Identification of likely buyers of specific products.
	Politics: Identification of fence-sitters (uncommitted voters).
	Banking: Identification of people with specific credit risks.
Analytics adaptation:	Collect personal data on height and weight from external data sources and compute BMI to identify overweight and obese individuals.
Analytics methods:	Predictive modeling, external data, integration.
Major challenges:	• Collection of personal data through external sources. • Effectiveness of intervention programs to reduce obesity.

This adaptation specifically focuses on the business need to reduce the burden of diabetes in terms of lives and dollars at stake. It homes in on obesity, a very serious epidemic and a major precursor of diabetes. The business goal is to reduce the incidence of diabetes through early identification of an individual's premorbid obesity, which is followed by targeted intervention programs to reverse weight gain. Many of these individuals will not have received sickness care, or might have received it for other reasons that did not include a risk assessment for diabetes. Therefore, there might not be any relevant data included in a traditional medical record that speaks to obesity. The analytics of multiple industries converge on this adaptation. Retail, banking, and politics have multiple data sources and sophisticated analytics to identify likely buyers, uncommitted voters, or probable defaulters, and they have programs to sell, convince, or reject those so identified.

The analytics adaptation is straightforward. The determination of the degree of obesity is easily determined by a simple calculation of the two variables of height and weight. The data are available from firms that aggregate publicly available data (and are not easily available from claims or medical records). The analytics methods include predictive modeling and the collection and integration of external data. The major challenges are belief systems that question the utility and appropriateness of external data and whether interventions intended to reverse obesity are effective.

Well-Being

The second analytics adaptation addresses the challenge to measure the health of populations in the way it is defined by the World Health Organization, that is, well-being (see Table 7.2).

Table 7.2 Analytics Adaptation #2

Population Health	Well-Being
Business need:	Activate people to co-produce health.
Business goal:	Reframe what success looks like from the customer's perspective to improve trust, exchange of information, and outcomes.
Industry sweet spot:	Politics: Polling.
Analytics adaptation:	Use BRFSS instruments, methodology, and oversampling of selected populations on specified measures to describe, monitor, and report on well-being.
Analytics methods:	Modeling, external data, integration, dashboards.
Major challenges:	• Perceived mission creep (from healthcare to well-being). • Unfamiliarity with external public data.

The measurement of well-being can have implications about the structure and intent of delivery systems because what is measured tends to get managed. This will facilitate stretching the historical boundaries of sick care to include total health. In addition to influencing the definition of what constitutes success, this adaptation has a direct purpose to measure what is important to people and thereby to earn their interest and engagement.

The business need is to activate people to co-produce health. The business goal is to reframe what success looks like from the customer's perspective to improve trust, exchange of information, and outcomes. The industry sweet spot comes from politics. Polling during presidential elections has been very effective in eliciting lots of pertinent data from voters.

The analytics adaptation is to use polling, or surveying, to determine well-being for a health plan, delivery system of municipality using an established, credible, and cost-effective approach. The suggestion is to use the Behavioral Risk Factor Surveillance System (BRFSS) on selected measures, oversampling for the specific population of interest. The data collection and analysis is mostly outsourced to the CDC. The analytics methods for a health plan or delivery system are mostly in the reporting, through customer-centric dashboards and the coupling of the information with other analytics adaptations, including social media.

The challenges include beliefs that there might be mission creep if healthcare expanded its purpose to address well-being. Also, there might be an unfamiliarity and discomfort with using publicly available databases.

My Dashboard

The third analytics adaptation also addresses the business need to activate people to be contributing partners in co-producing their own health (see Table 7.3).

Table 7.3 Analytics Adaptation #3

Population Health	My Dashboard
Business need:	Activate people to co-produce health.
Business goal:	Empower people to master their own self-care through the provision of compelling information, tools, and active support.
Industry sweet spot:	Politics: Dashboard.
Analytics adaptation:	Develop individualized "health journey" dashboards including key indicators, care plans, progress, and motivational and educational content; using expanded and tailored data; super-user-friendly tools; enabled through mobile.
Analytics methods:	Modeling, data collection, integration, dashboard.
Major challenges:	• Perceptions about (lack of) member engagement and use of prior tools such as portals and personal health records (PHRs). • Collaboration across organizational functions.

The business goal is to excite and empower people to master their own self-care through the provision of compelling tools and active support. This addresses a blind spot related to the paradox that clinicians need well-designed clinical support systems to provide optimal care, but patients do not and are essentially relegated to the role of recipients, but not participants, in using information to make important decisions about their care. The industry sweet spot, again, is politics. The dashboard developed and used by the winning presidential candidate in 2012 was lauded as a major factor in empowering campaign workers to persuade voters and get them to vote. This, in combination with other data-driven approaches, was the major factor in winning the election.

The analytics adaptation is an individualized "health journey" dashboard including key indicators, care plans, progress over time, decision frameworks, and motivational and educational content. It would use expanded and tailored data, use super-user-friendly tools, and be delivered through customer-preferred channels, like mobile.

The major challenges are the beliefs that member engagement through information products has been tried and failed before, including patient portals and personal health records. The existing design of these tools might not be up to the challenge. For example, some health-plan member portals are focused on getting members to fill out administrative forms more efficiently. This is not an optimal way to engage them, it underutilizes the technology, and it misses an opportunity to gain member trust. Another challenge is that the design features of such an expanded approach to a patient-centric dashboard involves a good deal of coordination and data integration across the enterprise and requires innovative new approaches to providing compelling presentations to consumers.

Radical Personalization

The fourth analytics adaptation also focuses on the business need to activate people to co-produce their own health (see Table 7.4).

Table 7.4 Analytics Adaptation #4

Patient Engagement	Radical Personalization
Business need:	Activate people to co-produce health.
Business goal:	Capture customer mindshare and motivate behavior change through radical personalization.
Industry sweet spot:	Retail: Segmentation, messaging, omni-channel, intervention.
Analytics adaptation:	Refine microsegmentation to the (nearly) individual level that aligns targeted interventions, delivered through customized messaging and channel preferences, to achieve better compliance for distinct clinical needs.
Analytics methods:	Predictive modeling, external data, integration, multichannel platforms.
Major challenges:	• Use of external data to build models. • Changing the focus of marketing customer analytics from selling products to improving health behaviors. • Belief that patients will not improve self-care.

The business goal focuses on capturing customer mindshare and motivating behavior change through radical customization. The industry sweet spot is from retail and it's well developed. The model from retail is the integration of market segmentation, messaging, omni-channel communications, and focused interventions.

The analytics adaptation is to refine microsegmentation to the (nearly) individual level and leverage the needs and predilections of each cluster with targeted interventions, matched with customized messaging and channel preferences, to achieve better compliance for distinct clinical needs. An area that is particularly ripe for this approach is patient compliance/adherence with treatment orders and medication prescriptions. The primary analytics method is deep microsegmentation based on significantly expanded and diverse data obtained from external sources. Predictive modeling is needed to determine the microsegment and its impact on the dependent variable, including its interaction with messaging, channel, and intervention.

The major challenges are the "usual suspects" in this area of activating people, including the use of external data and the belief that people cannot be activated to improve their own outcomes. Also, although the approach has been well-honed by marketing customer analytics to sell products, it may be a counterintuitive leap to use it to improve health.

Capabilities and Supports

The fifth analytics adaptation addresses the challenge to improve patient engagement by supporting and activating patients (see Table 7.5).

Table 7.5 Analytics Adaptation #5

Patient Engagement	Capabilities and Supports
Business need:	Minimize barriers to optimizing treatment outcomes.
Business goal:	Improve outcomes by addressing patient's challenges, capabilities, and needs to fulfill care plan expectations.
Industry sweet spot:	Banking: Credit risk assessment.
Analytics adaptation:	Develop and deliver Health Improvement Capability Score(s) to members of care team to support the focus on and attention to nonmedical determinants of health.
Analytics methods:	Predictive modeling, external data, integration, reporting.
Major challenges:	• Belief that nonmedical factors are outside of purview of healthcare and not modifiable. • Patients are difficult to activate toward self-management.

The overall business need is to improve outcomes by overcoming nonmedical barriers to successful care. The business goal is to improve outcomes by addressing the patient's challenges, capabilities, and needs to fulfill care-plan expectations.

The healthare analytics adaptation is derived from the banking and credit industries, which have refined the measurement of risk through the use of a broad array of data and sophisticated predictive modeling. The analytics adaptation is the Health Improvement Capability Score (HICS). It is a composite score of the overall capability of a patient to succeed with treatment based on the individual's level of activation, self-care skills, and the social determinants affecting the individual. The components of the overall score are also calculated. The scores would be made available to clinicians in the patient's medical record and to others who coordinate linkages with nonmedical services. The HICS scores might be as important as laboratory values in setting a course of successful treatment. The analytics methods include predictive modeling to determine the key factors related to poor outcomes, the collection of new data to populate the models, data integration, and clinical decision support reporting. The analytics capability to deliver is mature in general but variable by organization. The major analytical challenges include the capacity to build the new models and develop processes to collect new data from outside the industry.

The major challenges in the adoption of the overall innovation (which the analytics adaptation supports) is the perception that nonmedical factors are outside the purview of healthcare, that they are not modifiable, and that patients are generally disengaged from their own self-care.

Peer-to-Peer Provider Performance Ratings

The sixth analytics adaptation addresses the business need to improve provider performance through measurement and transparency, but with a twist (see Table 7.6).

Table 7.6 Analytics Adaptation #6

Provider Performance	Peer-to-Peer Provider Performance Ratings
Business need:	Improve provider performance through measurement and transparency.
Business goal:	Facilitate customer communications on their experience of care.
Industry sweet spot:	Politics: Social media.
	Retail: Social media.
Analytics adaptation:	Develop social media information platforms, like TripAdvisor for healthcare, that propagate peer-to-peer information sharing about healthcare providers and payers. Use information for performance measurement and improvement.
Analytics methods:	Social media, mobile platforms, performance measurement.
Major challenges:	• Accepting bottom-up-determined performance evaluations. • Beyond sentiment analysis to proliferating exchanges.

The business goal is to facilitate the expansion of customer communications about their care experiences. The rationale is that people have not been invited in or involved in providing feedback about their care to any appreciable degree, and that the information produced by experts through the traditional performance measurement approach is uninterpretable and unhelpful to people. They do not use it to make decisions about their insurers or their care, and the attempts to support this goal have been insufficient.

The industry sweet spot, from both retail and politics, is the use of social media, but not in a surveillance way to monitor sentiment. People have changed the power dynamics in other industries through mobile and social media by getting the information they need, when they want it, from trusted peers, and less so from sellers (except for price information, which they can instantaneously shop around).

The analytics adaptation is to develop social media information platforms, like TripAdvisor for healthcare, that propagate peer-to-peer information sharing about healthcare providers and payers. Healthcare can harvest this information but it cannot and should not control it, given the strong preferences exhibited by customers in other industries to eschew information domination from sellers.

The analytics methods include an understanding of social media and mobile and how to facilitate their use among the crowd of customers. The major challenges include existing beliefs about the nature of performance measurement, uncertainty about the benefits of social media, and relying on customers to lead the way on using information networks.

Team-Centered Outcomes

The seventh analytics adaptation addresses the business need to improve provider performance (see Table 7.7).

Table 7.7 Analytics Adaptation #7

Provider Performance	Team-Centered Outcomes
Business need:	Improve outcomes through performance measurement and improvement.
Business goal:	Improve outcomes by focusing on team performance.
Industry sweet spot:	Sports: Player and team performance ratings.
Analytics adaptation:	Demonstrate the ability to measure clinical performance at the care team level and its superiority in driving performance improvements relative to organizational levels of aggregation, such as hospital or medical group.
Analytics methods:	Predictive modeling, external data collection, integration, dashboards.
Major challenges:	• Resistance to performance measurement at a level more granular than the health plan, hospital, or medical group. • Belief that individual doctors are the key ingredient in patient outcomes.

The business goal is to improve outcomes by focusing on team performance. The industry sweet spot comes from sports, where performance measurement of individual players, along with full transparency, has been a mainstay of the industry.

One could focus on the paradox that consumers know much more about their sports heroes than about their doctors though they should know more about a doctor's "batting average." But we take a different view. Performance measurement of players has been more about entertainment than about winning games. The pressures on the business of sports have moved the measurement to outcomes (winning) and how the combination of player attributes and individual performance contribute to team wins. Indeed, the sports industry now realizes that after decades of concentrating on individual performance, the business needs to make teams work better.

The analytics adaptation is to develop the ability to measure clinical performance at the care team level and demonstrate its superiority in driving performance improvements and outcomes relative to existing organizational levels of aggregation, such as a hospital or medical group. The measurement should dovetail with the emerging reality that care, especially for the chronically ill, requires a well-functioning team of physicians, other caregivers, communities, and patients. The measurement might well influence the management of care to make the components work together for the good of the whole.

The analytic methods include predictive modeling, external data collection, integration, and dashboards. There are major challenges, including persistent resistance to performance measurement at a granular level and the belief that individual physicians are the key ingredient in producing patient outcomes. This might change the calculus about applying pay-for-performance incentives to teams rather than individual doctors.

Observations Across the Adaptations

There is a common thread running through six of the seven adaptations. Simply said, the way to improve outcomes is to ramp up the involvement of people in the co-production of their own health. There are many ways to improve health and healthcare and there are many stakeholders But the sweet spots from other industries do not provide insights that emphasize how to dramatically improve operations (although banking is expert in its mastery of transactions processing), or how to improve clinical practice (although healthcare has drawn on industry techniques for quality control), or how to work with government (although banking does have painful experience with this).

What is obvious across the industries is the laser focus on customers, to win them over to buy certain products, to convince them to vote for certain candidates, or to excite them to come to the stadium to watch the game. As industries emerge from the Great Recession, the linchpin for innovation and the transformations needed to achieve profitable growth always center on the customer.

Not so in healthcare...for a variety of reasons, including perverse payment incentives, insurance, and the highly professionalized medicalization of the business, which favors a directive rather than a collaborative relationship with patients. As a result of these defining characteristics of the industry, there is a huge blind spot. Healthcare has not seen or valued the sophisticated advances developed in other industries to engage the customer. But the engagement needed in healthcare is not an easy translation from other industries; for example, to sell the product.

Healthcare actually needs to change the current way it sells services through the fee-for-service system because far too many services are pure waste. Healthcare needs a drastic reorientation to improve outcomes. If it embraces that idea, the most important contributor to achieving better outcomes resides with the individual, not the medical system, community systems, research advances, and everything else external to the individual. The one big idea healthcare can learn from other industries is to take care of the customer. Paradoxically, healthcare should have the best relationships with its customers because of the nature of the caring profession. But it has been resistant to fundamental ways to partner with people; for example, by truly making care patient-centric and by sharing decisions with patients on many issues that are just not medical and are mostly about the preferences and tradeoffs concerning one's health.

The adaptations described in this chapter position analytics, in both simple and advanced forms, as potent catalysts to achieve these very important health goals.[1]

8

A Framework for Adopting Innovations

This chapter provides a framework and tools for progressing innovation from its early beginnings as an idea to its successful implementation into the operations of an organization. The Innovation Pathway describes the six stages of this journey. Next, we'll concentrate on the adoption of innovations and propose an Innovation Adoption Factors Model that describes the six areas or domains that need to be considered and mastered in order to persuade individuals to make the decision to adopt the innovation. We'll then break down the six domains into 18 factors and describe each of them. The factors become the basis for an adoption guidebook that can be used for evaluation and planning purposes. The chapter concludes with an application of the model to one of the healthcare adaptations proposed in the book through a hypothetical case study.

The Innovation Pathway

The pathway from an idea to its ultimate realization as an operable innovation is illustrated in the Innovation Pathway (see Figure 8.1). The pathway is made up of six stages, including idea, design, decision, execution, results, and reinvention. The stages are mostly linear, with the success of a given stage conditional to the passage to the next one. Failure at any stage extrudes the innovation from further development.

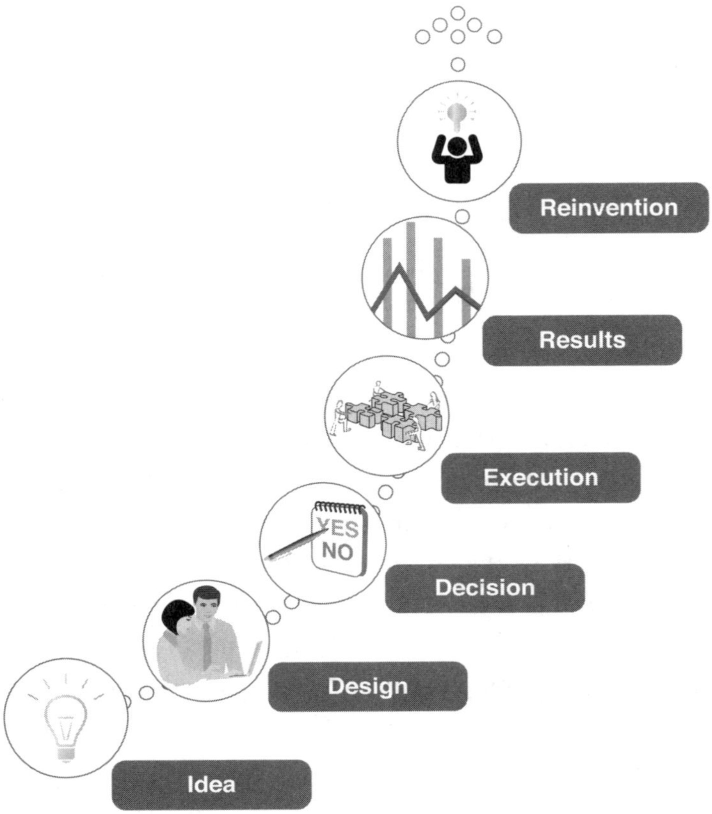

Figure 8.1 Innovation Pathway to Success.

The starting point of the pathway is an idea. President John F. Kennedy exalted ideas. He said, "A man may die, nations may rise and fall, but an idea lives on." Ideas can be many things, such as new knowledge that emerges from research, creative insights that come from thinking out of the usual mental models, or the collective wisdom of people working together to solve a problem. Ideas are often the backbone of innovations.

Ideas need to be converted into a plan of action during the design stage. This is the time to make the case that the innovation's "theory of action" will accomplish intended results. In healthcare, the end result is usually framed in terms of saving lives and dollars through the actions of providers, payers, and life science companies. The innovations can include prevention or treatment guidelines, ways to change patient behaviors, payment and delivery system reforms, and more.

The design stage involves decisions of individuals within an organization to adopt the innovation; that is, to accept or reject the plan. This stage requires facts, arguments, understanding of social and organizational context, communications to test the ideas and the plans, and last, but not least, leadership and management skills to bring the innovation over the finish line. Most ideas do not pass through the adoption membrane. Machiavelli said, "There is nothing more difficult to plan, more doubtful of success, nor more dangerous to manage than the creation of a new order of things." In healthcare it is frustrating that even programs of demonstrated worth take a long time to be adopted, if ever.

The next stage is execution or implementation. Goethe noted in the eighteenth century, "To put your ideas into action is the most difficult thing in the world." Implementation is certainly about following the design rule book in carrying out all the required process steps. It includes the need to fit the design prescription to the local needs of the organization, and this can involve a good deal of adaptation. And underlying it all, Jeffrey Pressman and Aaron Wildavsky, in their landmark book, *Implementation*, say, "Implementation is a struggle over the realization of ideas."

The next stage is results. Stakeholders need to understand performance in order to make adjustments in the design and operations so that it becomes institutionalized, altered to fit the changing needs of the organization, or discontinued. More often than not, ideas get stuck in the pathway and do live to fight another stage. And implementations of complex programs suffer from innumerable snags in delivery of the program and most fail.

The last stage is reinvention. Reinvention is important and is not a failure of intent to implement a plan "as written." For example, one might think that clinicians would adopt clinical guidelines, developed by their professional organizations, as is. But success might depend on not getting the full loaf and instead adopting a local version that works in a particular context. Hopefully, all innovations evolve and change through a learning and improvement process. The conclusion of Pressman and Wildavsky's book on implementation speaks to an ongoing cycle of innovation. They said, "Implementation is no longer solely about getting what you once wanted but, instead, it is about what you have since learned to prefer until, of course, you change your mind again."[1]

Innovation Adoption Factors Model

The Innovation Adoption Factors Model (see Figure 8.2) is a high-level map of the six domains that influence the likelihood of adoption of an idea or a design. Adoption is considered at the Decision stage of the Innovation Pathway. These six domains are composed of 18 specific factors, which are described in the next section.

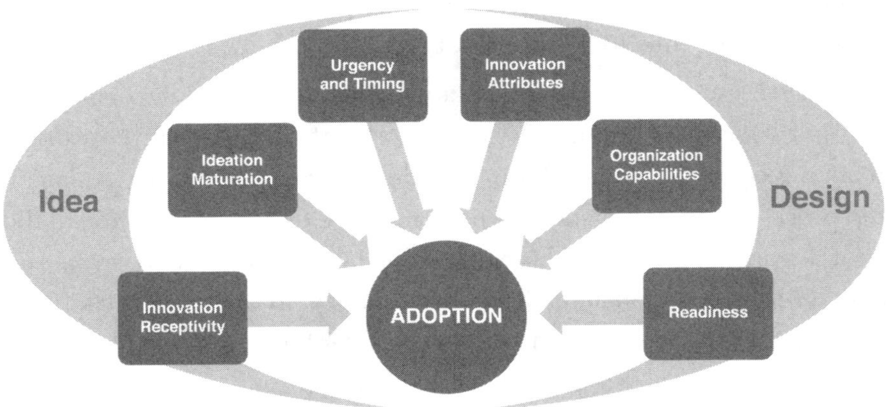

Figure 8.2 Innovation Adoption Factors Model.

Note that the overall model and its domains and factors are adapted from the extensive work of a number of researchers and policy experts in the innovation diffusion field, including Everett Rogers on diffusion theory,[2] William Eggers and John O'Leary on getting things done in government,[3] John Kingdon on agenda setting,[4] Yeheskel Hasenfeld and Thomas Brock on policy implementation,[5] Pressman and Wildavsky on program implementation,[6] Malcolm Gladwell on the tipping point model of change,[7] and Paul Plsek on complexity theory.[8]

Adoption is about making a decision to activate an innovation into practice. The process of adoption is largely about collecting, processing, and evaluating information to understand the innovation and to reduce uncertainty in relation to its pros and cons.

The model is composed of two halves. On the left side is the idea stage and on the right is the design stage, each with its respective factor domains. At the successful completion of these two stages, meaning that the 18 factors have been addressed, a recommendation for a decision can be considered by potential adopters.

Adoption Domains and Factors

This section provides a description of each of the 18 adoption factors. The factors are presented as rating checklists that can be used in evaluating the adoptability of the Key Healthcare Adaptations described in the previous chapter for one's own organization. A case study at the end of this chapter provides an example of using the factors for decision making.

Idea Stage

The idea stage includes the domains of innovation receptivity, ideation maturation, and urgency/timing (see Figure 8.3).

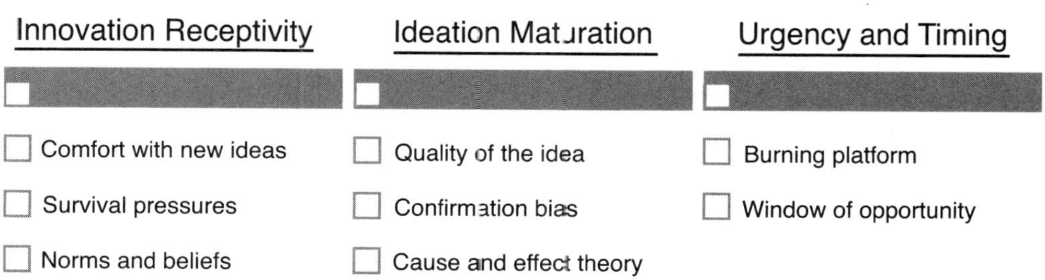

Figure 8.3 Idea generation domains and factors.

Innovation Receptivity

Innovation receptivity includes the factors of the organization's comfort with dealing with new ideas, its survival pressures, and its existing norms and beliefs.

Comfort with New Ideas

Some organizations thrive with change and some shrink from it. Andrew Pettigrew, Ewan Ferle, and Lorna McKee, in their book *Shaping Strategic Change: Making Change in Large Organizations*, coined the phrase "receptive context" to describe the degree to which an organization is comfortable with change and new ideas.[9] Organizations with high receptive context are continually "ripe" for change and adopt new innovations relatively quickly. Other organizations with low receptive context, which are confronted with the same challenges and are aware of the innovation in the same way, lack the will or ability to implement the innovation. This is really the metaphorical case of whether the land is fertile, arable, and moist enough to grow the seed.

Don Berwick, the founder of the Institute for Healthcare Improvement, suggests that there are cultures that can encourage and support or discourage and impede change.[10] For example, a nurturing environment would reinforce and reward the work of "boat-rocking" innovators and offer them security for the inevitable failures that are part of the work of experimentation. A penalizing environment takes issue with these "troublemakers" and spreads the message that others should do so as well, resulting in a marginalization of this key talent.

Survival Pressures

Innovations take energy and can be an opportunity cost that displaces other more important organizational priorities. For example, a health plan that is devoting lots of energy to compliance issues of the Affordable Care Act (ACA) or to replacing a legacy system data warehouse might not have the patience or tolerance for new ideas. Similarly, an organization that is just getting by financially and has limited resources might not be able to invest time and money into the consideration of innovations. Of course, organizations that avoid innovation because of chronic survival pressures might set themselves up for a death spiral of diminishing capability to innovate. The consequences of inaction could become very clear at some point, and this awareness might rearrange priorities for reconsideration of the innovation.

Norms and Beliefs about Change in General

Existing norms and beliefs of individuals, organizations, and industries affect how innovation is regarded. Healthcare is a conservative industry due in part to some ingrained beliefs from medicine. For example, the Hippocratic oath includes the famous phrase *primum non nocere*, or "first do no harm." To this day, this is a fundamental principle for medical services. The twenty-first-century translation can be stated as "given an existing problem, it may be better not to do something, or even to do nothing, than to risk causing more harm than good."[11] Similarly the phrase "It's better to be safe than sorry" reinforces the inclination to be cautious and to fend off legal suits. These beliefs can lead to an overestimation of the risks involved with new ideas and innovations and result in stifling them. Similarly, doctors are very autonomous professionals whose norms include self-reliance and resistance to imposed ideas.

Ideation Maturation

The ideation maturation factor is about developing a sound idea and includes the quality of the idea, resolving confirmation bias, and demonstrating a clear cause-and-effect theory of how the innovation produces change.

Quality of the Idea

The idea needs to be well formulated It should emerge from good theory, be evidence based, supported by converging ideas in the literature, and logically developed with good critical thinking. If it passes these hurdles, it enters into what Kingdon refers to as the "primordial soup" of ideas and is ready for further debate to eventually be considered when just the right combination of circumstances arises.

A new, innovative idea involves a creative thinking process. Involving a wide circle of individuals in the creative process, and especially those who will be involved in deciding on and implementing the innovation, is critical to the maturity of the idea. Sometimes the best ideas come from out of the blue, for example, from another industry, and the initial intrigue needs to be supported by its logical linkage to healthcare. Sometimes they come from the deliberate work of science and research. In this case, a need/problem is recognized, hypotheses are formulated to test it, and solutions emerge from the findings. However the idea emerges, the quality of it sets the stage for everything else that follows in the Innovation Pathway.

Confirmation Bias

Eggers and O'Leary refer to the seven deadly traps that can bedevil any important change effort. The first is the Tolstoy Trap. They refer to this as the notion, attributed to Tolstoy, that "we see only what we are looking for often by staying blind to what is in front of us."[12] This is also referred to in the research literature as confirmation bias, which is a cognitive bias in which people seek out information that supports their beliefs and ignore all the rest. It is a very powerful bias, possibly related to a basic human need to be right and avoid embarrassment, as well as to avoid conflict. However, it is a big threat to critical thinking and the logical development and full expression of ideas. The antidote, the way to resolve the trap, is to constantly and actively challenge one's thinking, to be humble about one's knowledge, and to invite all voices, especially contrary ones, to offer a point of view and evidence. Also, it is a wise strategy to engage people in idea generation because there is a preference to favor ideas that one has been involved in generating.

Cause and Effect Theory

The idea must be such that the innovation (X) will lead to the intended results (Y). This is about the potency of theory and the model and not about the implementation of the innovation, which is quite another worrisome matter in terms of achieving Y. The steps must be clearly articulated and lead eventually to clear technical specifications and details on how to carry them out. When there is technical uncertainty about cause and effect, other factors can dominate the discourse about the innovation, such as ideological and political issues.[13]

Urgency and Timing

The urgency and timing domain addresses whether there is a burning platform for change that the idea is aligned with and that there is a window of opportunity within which action is possible.

Burning Platform

Organizations face many demands and give attention to the highest priorities. Getting on the radar screen can be difficult for a new idea, depending on its potency and the status of factors discussed earlier; for example, receptive context. A burning platform provides compelling information that the status quo is dire or changing dramatically and that the current ways of doing business are not up to the challenge and a new solution is needed. If a burning platform is not established, it might be impossible to get the attention of key decision makers regardless of the merit of the proposed innovation. If attention is gained, the innovation must be shown to align with the burning platform.

Window of Opportunity

Kingdon refers to the window of opportunity as a time period during which change can occur and ideas can be considered. For example, in the political context this can be the first year of a President's second term. In healthcare, a time frame might be imposed by a regulatory requirement, or a change in leadership, or funds that become available at a certain time. The window of opportunity is more than a time frame. It is a coupling of the burning platform, the innovation under consideration, and the right "moment." In other words, when the iron is hot, bend it.

Design Stage

The design stage includes the domains of innovation attributes, organization capabilities, and readiness to position the innovation for adoption (see Figure 8.4).

Innovation Attributes	Organization Capabilities	Readiness
☐ Relative advantage	☐ Individual's acceptance	☐ Prototype
☐ Compatibility	☐ Program theory/model	☐ Communications
☐ Complexity	☐ Skills and competencies	☐ Leadership
☐ Trialability		

Figure 8.4 Design domains and factors.

Innovation Attributes

The innovation attributes domain is very important and includes the key factors of the relative advantage of the innovation, its compatibility with existing processes and attitudes, its complexity, and its "trialability."

Relative Advantage

Relative advantage is the degree to which the innovation is judged to be better than an alternative or the status quo. It usually addresses the measurement of the potential impact in financial terms, such as the return on investment. In healthcare it often addresses the impact of the innovation on health and healthcare with such measures as reductions in mortality, lessened disability, improved functioning, and decreased adverse events. It is considered to be the most important parameter for the acceptance of an innovation. Berwick states that the perceived benefit of an innovation predicts between 49% and 87% of the variance in the rate of spread of an innovation.[14] (It should be noted that the rate of spread is different from the rate of adoption, but the magnitude of the importance of relative advantage is large in both cases.) If idea generation answers the question, "what is it?," then relative advantage addresses the question, what is it going to do for my company?

Compatibility

Compatibility addresses the degree to which the innovation is perceived to be consistent with the ways of the adopter organization and individuals. It addresses these questions: Does the innovation fit in with the way we do things? Is it out of line with how we have done things in the past? Does it reflect our values? Is it aligned with an overarching strategic plan? Does it fit with existing information technology? Does it blend well with other programs; for example, does it support or hinder an associated program?

Complexity

This is the degree to which an innovation is perceived to be difficult to understand and use. The acronym KISS applies here and more generally to the selling and execution of any idea or product: keep it simple, stupid. Good marketing can simplify the idea of complex innovations. The innovation needs to be decomposed into the steps or processes to make it understandable. Overly complex designs might be difficult to comprehend by adopters and difficult to interpret for implementers. An antidote to saving on overly complex innovation is to adapt it in a simpler way without compromising its essential working parts.

Trialability

Trialability addresses whether the innovation can be tried and evaluated on a small-scale basis as a test before a full-scale rollout. This diminishes the potential risk of failure and subsequent embarrassment to the adopters. It provides more evidence on the potential worth of the innovation to help sell it and complete the adoption of it. Most innovations can and perhaps should be sold as incremental stages with advancement to the next phase contingent on the results of learned experience to that point. However, there are risks to an incremental approach. For example, the results might not be statistically compelling on a small-scale test (small sample size), and the early phase of implementation might not have worked out all the kinks, whereas a longer period might be more indicative of the true functioning of the innovation.

Organization Capabilities

The organizational capabilities domain includes individuals' acceptance, whether the organization can deliver on the program theory, and whether the innovation is a complete or partial solution.

Individual's Acceptance

The decision to adopt an innovation is made by an individual or a group of individuals in an organization. For an analytic innovation, the decision can be made by a manager of purchasing or all the way up to the executive committee of the board, depending on the investment size, risk, and importance of the innovation. After an adoption decision is made, it is expected to be carried out by the staff. It becomes a management priority, tools and educational packets are provided, performance is monitored, and the intended results are expected.

However, the process is complicated. The success of an innovation is usually not determined by management fiat but by the decisions of many individuals on whether they want to change relative to their own values and needs and organizational politics. For example, a hospital administrator decides to buy a new application for the electronic medical records system that issues alerts when doctors do not follow guidelines for ordering imaging tests. However, individual physicians will decide whether to pay attention to the alerts and change their behavior. It could be that they have alert "fatigue" and regard the alerts as distractions not worthy of their attention. Putting teeth into physician compliance by the hospital administrator sending a "reminder" note or docking pay might lead to more resistance from the doctors, who might resent a top-down approach to change that threatens their autonomy.

The reality might be that healthcare is a complex adaptive system that does not respond or adapt well to a linear, rational, mechanistic, and command-and-control management approach. Plsek states that "a decision to change is ultimately made by individuals in a complex system according to personal mental models about such things as the benefits and risks associated with the change [for them]."[15] This factor highlights the need to include information about the likely acceptance of the innovation by individuals into the adoption decision. After all, if individuals are going to resist it with force, why go through all the trouble? When it comes to innovation, the engagement of individuals is more than "buy-in." It has to be genuine, be continuous throughout the Pathway process, and involve a high degree of communications.

Program Theory/Model

Two aspects of innovations are discussed in this book. The first is the analytics component. This can be a specific predictive model, the use of external data, or a new initiative to gather more information on customers. The analytics component supports the programmatic intervention. Both components of the intervention need to function well on their own and in synchrony with the other. An example is the Health Improvement Capability Score. This score provides vital information on a patient's capacity to respond to treatment in terms of their "activation" and the social forces that work against their getting better. The analytics solution is forthright and doable. It can be a huge success, but the intervention fails. Why?

The issue is whether the programmatic component can use/absorb the analytic contribution to achieve its goals. If there is not a sound programmatic model to effect a change, even with full accessibility to the analytics, the innovation will fail. Another example is the analytics contribution of BMI scores to identify people with premorbid risk of obesity and diabetes. The analytic task is, again, straightforward. But it is not at all clear whether the treatment team has a viable intervention model to reverse weight gain. It's possible that the availability of these analytic assets will spur programs to develop better capacity. But it might be futile to deploy sound analytics technology when there are deficits in the "sociology." If the model does not exist, is broken, and/or if there is not a pathway to improve it, then all the good analytics intentions and effort might be for naught.

Skills and Competencies

The organization must have people and technologies with the skills and competencies to carry out the analytic solution and the programmatic intervention. New innovations are likely to place new demands on individuals who are engaged with the adoption of the innovation. A good model and plan will not succeed if the right talent and technical supports are not available.

Readiness

The readiness domain includes evidence from small-scale tests of the innovation, communications to position the innovation favorably, and, last but not least, leadership.

Prototype

The trialability parameter previously described assesses whether the innovation can be tested on a small-scale basis. Prototyping involves the actual test-drive of the innovation. It involves building out the innovation design, or a portion of it, and testing it in relation to whether it is efficacious and implementable. It provides real-world evidence about the innovation and can be very influential in the adoption decision. Prototyping supports the notion of rapid experimentation to learn and adapt. And it roots out possible failure points by directly testing how the innovation might fail (better to "fail safe and fail fast"). It involves careful engineering approaches of flowcharting and process mapping to work out each step of the design. Prototyping facilitates what Rogers calls the "observability" of the innovation such that potential adopters can watch it in action and thus facilitate more discussion about it.

Communications

Rogers states that the diffusion process at its core is about communications. The communications are about new ideas and whether they are worthwhile to try. The communications are among people who share information in an effort to arrive at a mutual understanding. He asserts that most individuals do not decide to adopt an innovation based on facts and scientific studies but rather on the observations from other people (peers) who have tried it. Those who have been the first to try out an innovation are referred to as early adopters and can become opinion leaders who can have a great influence on peers. This suggests that ample interpersonal communications about the innovation make a positive contribution to the likelihood of adoption.

Communications can be facilitated through existing social networks that encourage the free exchange of ideas. Organizational roles that deploy communications well to convince individuals about the merits of an innovation include connectors, mavens, and salesmen, in Gladwell's terms.[16] A connector has many and varied relationships in the social system and can make introductions across boundaries and help to knock down barriers. Mavens like to spread ideas and act as information brokers who share ideas and help people solve problems. And salesmen are the "persuaders" who have a knack for making others agree with them.

Innovations can be packaged to make them "irresistible" to those considering them. And they can be packaged with the needed support materials such as documentation and user manual, to demonstrate that implementation concerns have been considered. Consulting firms are expert in creating "fashion consciousness" to increase demand for their engagement with clients; for example, the assertion that "big data" is the next wave.

Leadership

Leadership at the executive level is about creating the vision and executing a strategy to "get people from where they are to where they have not been," in the words of Henry Kissinger. Succeeding at innovation requires leadership that creates an environment that welcomes and encourages new ideas and change, cuts through the confusion and uncertainty to be decisive, and then leads the organization to execute flawlessly. Ideally, the organization is a learning system that is constantly updating its mental models and making improvements. Without this executive leadership, the rate of adoption of new innovations is compromised.

In addition to executive leadership, there needs to be "boots on the ground" leadership that is instrumental in moving a specific innovation through the process and concluding with a handoff at the end of the project to the responsible parties who will embed it in the daily life of the organization. These leaders are sometimes referred to as change agents and they can be internal and/or external. Their primary role is to be an excellent networker and communicator across the multiple groups involved in the process. They help individuals see the burning platform for change and understand organizational needs and capabilities, organization gaps, and possible future states. They facilitate the full discussion of ideas and the refinement of the innovation proposed for adoption. After the innovation is adopted, they can help keep it on course through the early stages of implementation by unlocking resistance or adapting the innovation to make it more workable.

Case Study

This case study uses the Factors Model to evaluate the likelihood of adopting an analytics innovation by a hypothetical organization. The evaluation results can be used for planning purposes to develop a road map for adoption success.

The case involves a health insurance company that is concerned about the poor treatment outcomes of its Medicaid population. It is considering the adoption of the Health Improvement Capabilities Index.

Context

Description of the Analytics Innovation: Health Improvement Capabilities Score

Table 8.1 outlines the key business issues and analytics challenges. The overall business need is to improve outcomes by overcoming nonmedical barriers to care. The business goal is to improve outcomes by addressing patients' challenges, capabilities, and needs to fulfill care-plan expectations.

Table 8.1 Case Study Evaluation: Capabilities and Supports Parameters

Patient Engagement	Capabilities and Supports
Business need:	Minimize barriers to optimizing treatment outcomes.
Business goal:	Improve outcomes by addressing patients' challenges, capabilities, and needs to fulfill care-plan expectations.
Industry sweet spot:	Banking: Credit risk assessment.
Analytics adaptation:	Develop and deliver Health Improvement Capability Score(s) (HICS) to members of care team to support the focus and attention to determinants of health.
Analytics methods:	Predictive modeling, external data, integration, reporting.
Major challenges:	• Belief that nonmedical factors are outside of the purview of healthcare and not modifiable. • Patients are difficult to activate toward self-management.

The healthcare analytics adaptation is derived from the banking and credit industries, which have refined the measurement of the risk of loan default. The analytics adaptation is the HICS. It is a composite score of the overall capability of a patient to succeed with treatment based on the individual's activation and self-care skills and on the social determinants affecting the individual. The components of the overall score are also calculated. The scores would be made available to clinical teams in the patient's medical record, as well as to other health support teams. The analytics methods include predictive modeling to determine the key factors related to poor outcomes, the collection of new data to populate the models, data integration, and clinical decision support reporting. The analytics capability is mature in most of these areas, with the analytical challenges including the capacity to build the new models and develop processes to collect new data. The major challenges in the adoption of the overall innovation is the perception that nonmedical factors are outside the purview of healthcare, that they are not modifiable, and that patients are generally disengaged from their own self-care.

Health Plan Context

The hypothetical health plan is 50 years old, regional, one million members, and marginally profitable. It has a 15% market share of the commercial market and a favorable reputation among employers. It is second to the Blues plan, which has a 60% share. It has a history of succeeding and failing as a staff model HMO. It has recently acquired a smaller Medicaid insurer to broaden its penetration into this expanding market. It is planning for partnerships with some provider groups to build an Accountable Care Organization pilot program. The new CEO envisions a future where the insurance component of the business will be complemented with new and more profitable business lines. The plan recently created a new position of Chief Analytics Officer (CAO), who will be responsible for leveraging its analytics assets to improve earnings by reducing medical costs and by selling medical information services. The states it serves are proceeding on schedule to develop their own health insurance exchanges and have accepted the ACA Medicaid plan for expanded enrollment.

Adoption Factor Evaluation

The readiness of the organization on the six domains is evaluated next. The dashboards include a rating on each of the 18 factors using the standard green, yellow, and red convention to indicate good, medium, or poor readiness for the factor. The component factors are combined into an overall adoption decision index. The component and index scores are displayed in an adoption decision dashboard.

The first three domains are for the idea generation stage.

Innovation receptivity scoring is displayed in Figure 8.5a.

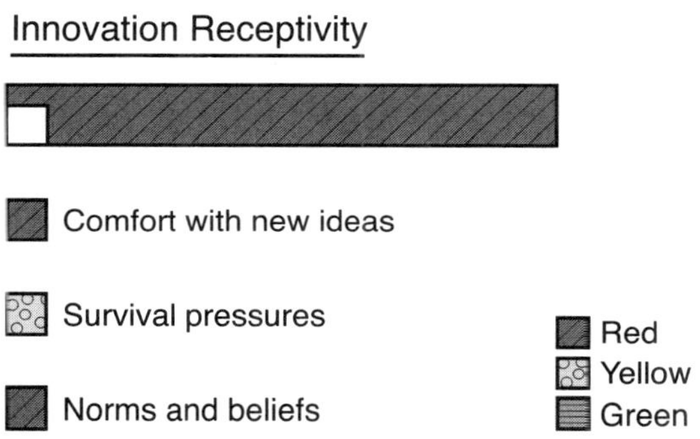

Figure 8.5a Case study evaluation: idea stage, innovation receptivity.

The organization's innovation receptivity is low overall, clearly in the red zone, and a significant barrier to change.

Its comfort with new ideas is in the red zone. The departing CEO held the position for 15 years and the culture of not rewarding risk taking is ingrained. The organizational structure is hierarchical and communications up and down and across it are very formal. New ideas are considered only during the six-week yearly budget cycle and the process is arduous and stern. Staff who rock the boat are considered troublemakers and their motives are questioned.

The analytics functions are dispersed throughout the organization, with much of the power and budget centralized in IT. There is a small medical informatics function that provides thought leadership. The few departments and functions that produce their own analytics are autonomous.

The organization has worrisome survival pressures. It failed as an HMO and merged with an indemnity insurer, and this history reinforces its reluctance to innovate and risk failure again. It has many compliance requirements from the ACA, as well as new state regulations. Its markets will be very active in the health insurance exchanges, and it must transform dramatically to become more customer-centric. The state has regulated that all plans serving state employees and residents enrolled in state insurance plans (commercial and Medicaid) must shift from fee-for-service payments to global payments. The state is also actively encouraging the development of Accountable Care Organizations (ACOs) through regulation and incentives. All of these requirements placed new demands on the IT function such that it had to change its legacy data warehouse. A major, multimillion-dollar project is underway to do so.

A potentially positive development is that a new CEO was hired to radically reorient the organization to respond to its new opportunities and threats, including a reorg to flatten the organization. How successful the management changes will be in affecting the long-standing cultural predilection to shun change is unknown. History would suggest a red rating but the hope in new leadership edges it up to a yellow.

The norms and beliefs of the organization are conservative. As with all insurance companies, its core strength is in assessing risk. However, this strength has a flip side. It is also an impediment when it comes to thinking outside the box, which requires a high degree of comfort with the risks involved in such change. The organization is severely siloed into layers and functions, and it is hard to modify individual positions and socialize a collective perspective. Change is driven from the top and staff are expected to comply.

Were it not for the significant survival pressures, the hope that a new CEO will make a difference, and the external demands, *any* proposal for change at this time would be considered dead on arrival.

Idea maturation (see Figure 8.5b) is in the yellow zone overall.

The HICS idea has been well formulated and has been reasonably socialized. The leadership staff understand that patients need to be activated and supported in order to achieve better outcomes and that doctors need to be more attuned to factors that can compromise their medical intentions. There is agreement that a metric will help raise awareness and spur the development of programs to fill gaps.

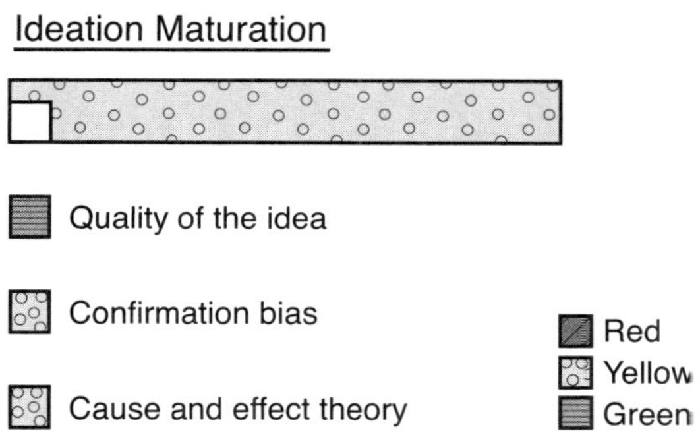

Figure 8.5b Case study evaluation: idea stage, ideation maturation.

In terms of confirmation bias, the new CAO is driving the idea. She has a strong bias toward population health. But some individuals do not get it, believe that it is a not an insurance function, and think it is far-fetched and in the category of "boiling the ocean." Hence, they have not been predisposed to discuss the idea on its merits. The actuaries are a powerful voice in the analytics community and in the insurance business overall, and they have ignored the idea. The paradox is that the idea is all about understanding risk, which is at the core of their work.

The cause-and-effect theory is understood, notwithstanding the resistance previously mentioned. It is clear that social determinants and patient activation have an enormous influence on outcomes, and that it is possible to measure these correctly. What plagues the acceptance of cause and effect is whether the health plan can and should do anything about it. Is it the health plan's role and responsibility to improve outcomes? If so, is it technically possible to change behavior and social conditions sufficiently to achieve some potency with the innovation?

The overall score for urgency and timing (see Figure 8.5c) is a solid green.

The burning platform is compelling and might ignite progress on other, lagging factors. A major driver is that Medicaid plans, state employee insurance plans, and ACOs (eventually) will be paid through global payments (also known as neo-capitation). This will reward health plans to lower medical expenses. If the actual medical experience is lower than the projected expenses reflected in premiums, the plan and its partners reap the profits. Also, if the plan can lower premiums due to its success in lowering costs, it will attract a larger market share. The states will also have a quality performance measurement system to ensure and pay for high quality. The ACA also provides grants through the CMS Innovation Center to improve outcomes, and this innovation qualifies.

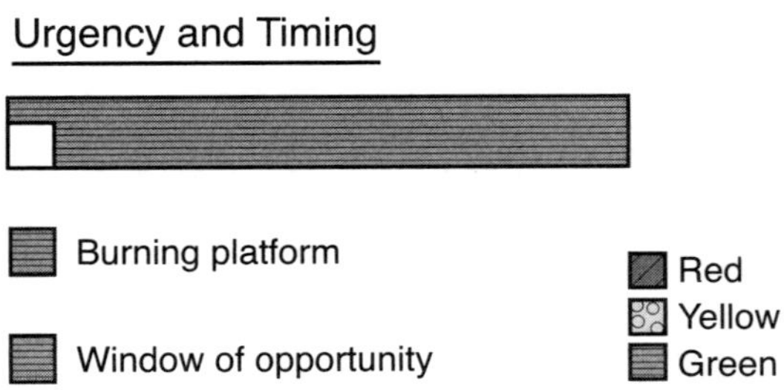

Figure 8.5c Case study evaluation: idea stage, urgency and timing.

Everyone understands that 40% of health functioning is attributable to personal behaviors and 15% to social factors. Even if the healthcare delivery system were to work perfectly, it would have a relatively modest effect on outcomes. The other behavioral and social levers are needed to optimize outcomes. The HICS is aligned with this need. And it is recognized that the HICS might also stimulate partnerships with provider groups and nonhealthcare organizations to create solutions that work.

The window of opportunity is also favorable for adopting the innovation. The health exchanges enroll members in the fall of 2013. The states require global payments in 2014. CMS grant proposals are due in six months. And the new CEO and CAO are looking for ways to provide value to the organization in a short time frame to demonstrate their value and retain their positions. The clock is ticking for action.

Next is a review of the design stage factors. The first is the important area of the innovation attributes (see Figure 8.6a). Overall, the innovation attributes are in the yellow zone.

An important reminder is that the innovation package is composed of two elements. The first is the analytics adaptation; that is, the HICS. The other is the program intervention. They are inextricably bound together, and both must function well independently and in synchrony to produce the intended results. In this section, the analytics adaptation is reviewed in terms of its innovation attributes. The program component is addressed in the organizational capacity domain.

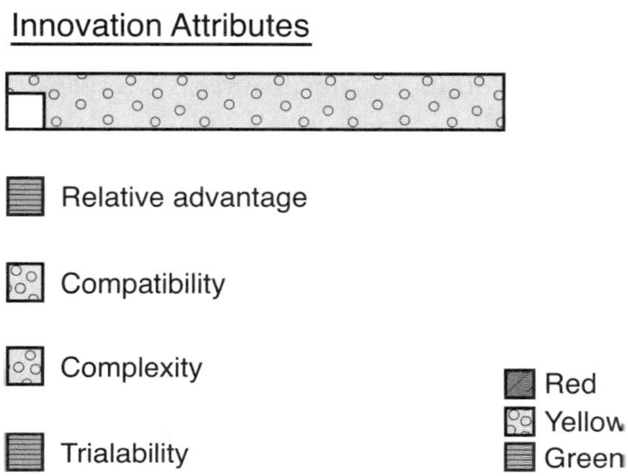

Figure 8.6a Case study evaluation: design stage, innovation attributes.

The HICS is unique and there is no other option quite like it in healthcare today. So the determination of the relative value uses the status quo (without a score) as the comparator. A detailed staff study of the literature demonstrated that the potential for cost savings for the Medicaid population deriving from improvements in patient adherence to treatment plans—including prescription medications; patient behavior modification to reduce smoking, drinking, obesity, and lack of exercise; patient activation to improve health monitoring and social connectedness; and social supports to improve access to good foods, transportation, insurance, and legal aid—is significant and on the order of 33% per year. There are few studies that provide conclusive evidence on the actual effectiveness of interventions that have been put into practice. The consensus was that a reasonable estimate of savings could approach 20% per year with full implementation and benefits of the programs phased in over a five-year period.

The study concluded that the analytics component, the HICS, is a necessary foundation for the success of the intervention because it gives guidance to the physician and health team on how to direct their efforts. The report emphasized that what is measured is managed. They concluded that without the HICS, but with some education, the savings would diminish to less than 5%.

In terms of return on investment, the costs of developing the HICS include model building and the determination of the effects of different independent variables on outcomes and costs. The biggest challenge is collecting the data in the most cost-efficient way. One approach is to collect the relevant information directly from the patient as part of the initial workup and at regular intervals thereafter, not unlike the collection of clinical history information. The clinical history process gets reframed as a total health history, including patient capabilities. This could be administered through a self-service Internet application or an interview by an associate health worker. An alternative data source is from public records and is available from private data aggregators. The initial recommendation was to evaluate the publicly available data and test it with the models on historical data. This data-capture approach was judged to have lower impact on clinical operations and lower cost. The ongoing costs of these data were estimated at $0.50 per member. Costs of storing and integrating the data with claims data was estimated at $0.05 per member. Initial costs for predictive modeling was estimated at $250,000. Overall, the ROI was considered sufficient at a 10% return per year and it was determined that a pilot test on a small sample of historical patients should be undertaken.

The compatibility of the HICS with the existing ways of the organization warranted a yellow signal at best because the idea is outside the realm of the usual insurance box and the data source is outside the comfort zone of standard claims and member data repositories. Additionally, the innovation would have to be implemented within a delivery system that might be harder to manage than within the health plan. (The health plan is actively looking for provider partners in order to have greater leverage on medical management in general.) A positive indicator is that the marketing department became very interested in the prospect of using the newly acquired data to improve its 360-degree view of the customer and could be useful in more refined approaches to microsegmentation.

Complexity also received a yellow signal because of the unknowns related to the new research and new data sources needed to make the HICS work. The HICS is not a difficult concept to communicate to providers. The analytics staff provided demos of what it would look like in the medical record in an attempt to simplify the understanding of the idea.

Trialability was feasible given the suggestion to first test the intervention on a historical sample. The suggestion was made to do a pilot test on one of the Medicaid diabetes programs, and initial conversations with the state Medicaid office were encouraging about their support for the innovation.

The organizational capabilities (see Figure 8.6b) to absorb and make full benefit of the innovation are in the red zone and are a major barrier to a favorable adoption decision.

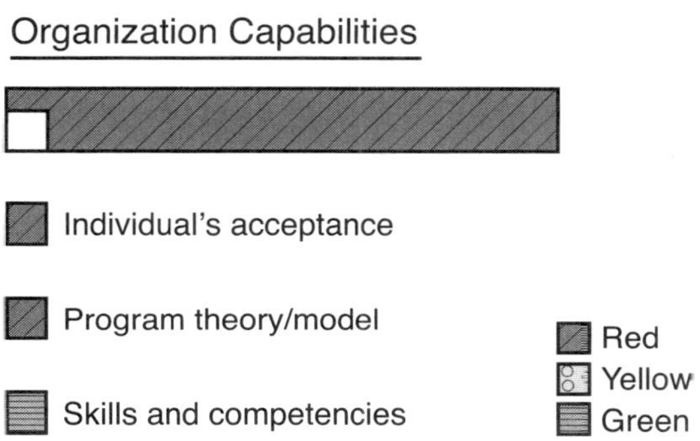

Figure 8.6b Case study evaluation: design stage, organization capabilities.

In terms of an individual's acceptance, the biggest concern is that physicians might not regard the innovation positively or support or use it. The emerging consensus is that it might not be the physician's job to address nonmedical issues, and they should keep to the knitting of medical care and work with others on the multidisciplinary team to coordinate the use of the HICS. Ideally, there would be a team that evaluates all the relevant information pertaining to factors that limit a successful course of care which then coordinates appropriate actions. The health plan expressed interest in the work of Health Leads (described in other chapters of the book). Health Leads would receive the HICS on all chronic-disease patients and follow up with the patient as needed to get more information. They would develop a support plan, inform the medical team, and coordinate with other organizations in the social services network to attend to specific needs. However, the Health Leads concept is somewhat foreign to the usual delivery of medical care, and its benefit has not been demonstrated through research, which might engender skepticism among some individuals.

The program theory is sound. However, the program model is exploratory and there is early but sparse evidence of its effectiveness. A study on the experience of Health Leads to lower medical costs is underway and will speak to this component of the innovation. Emerging evidence demonstrates ways to improve medication compliance. Patient activation models are demonstrating success in reducing costs.[17] And other ways to ramp up the level of patient engagement for self-care are emerging and hopeful. All of these findings were the basis of the consensus estimates of savings referred to earlier.

The skills and competencies for the analytic innovation are excellent. The new CAO is intent on demonstrating how analytics can improve earnings and sees this as a promising area. She is also considering the formation of an Analytics Center of Excellence, which will concentrate expertise from within and outside the company on fast-track, high-priority projects.

Readiness (see Figure 8.6c) overall is clearly in the green zone.

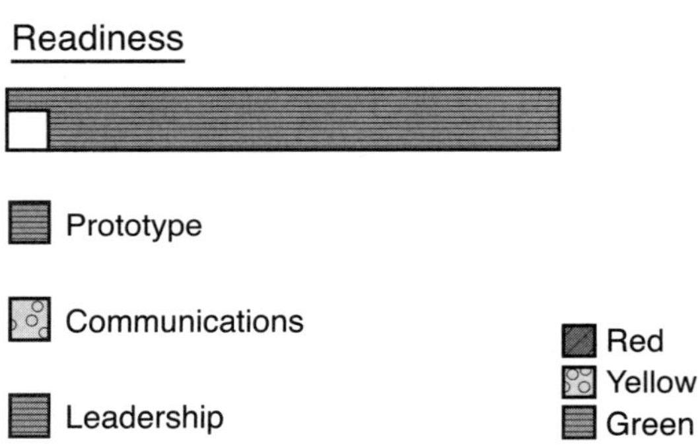

Figure 8.6c Case study evaluation: design stage, readiness.

A conceptual prototype of the HICS includes a full delineation of its functional and technical specifications. Preliminary modeling identified ten dimensions of capability based on a multiple regression equation using selected external data as the independent variables and annual medical costs as the dependent variable. The model could not test the effectiveness of the specific program interventions based on historical data, but it did sort the sample into subgroups based on prescription adherence, which was a proxy for high-, medium-, and low-activated patients.

The results confirmed the magnitude of possible savings when comparing the low-activation group with the high-activation group. The results were shared with focus groups of caregivers, and the feedback was used to improve the positioning of the HICS in communications.

Communications are a concern. The organization has been characterized as "top-down," including the one-way direction of communications and the command-and-control management style. Lateral communications have not developed sufficiently and most individuals would rather eat alone at their desk than share food and ideas in an informal way with peers. The CAO is considering a new function of the center of excellence to build competence in research translation and the diffusion of innovations. Her first initiative in this regard is to identify opinion leaders, connectors, mavens, salesmen, and early adopters. She considers these skills as key success factors that represent analytic skills that are just as important as the technical expertise needed to build data warehouses, do computations, and provide reports. The CAO is clearly the change agent for the innovation.

The new leadership recognizes the imperative to innovate and to change the culture to encourage and reward change, new ideas, and risk takers. The CEO believes that the future of the health plan is in partnerships with providers and other community organizations to co-produce health and reduce costs. He also believes that one of the organization's core skills is analytics and that it can drive the business toward needed transformation. In addition to the executive leadership, a change agent has been assigned the daily operational responsibility to get the HICS adopted and operating successfully.

The overall dashboard (see Figure 8.7) for the results obtained from the innovation adoption factors model indicate an overall cautious, yellow rating for the adoption decision index.

There is an equal balance among the six domains between red, yellow, and green, with the organizational receptivity to innovation and its capabilities to master change in the red. On the other hand, the timing and readiness to move forward at this time are clearly in the green. The areas in the yellow include the maturation and acceptance of the idea and the innovation attributes, both of which are very critical for the successful adoption of an innovation.

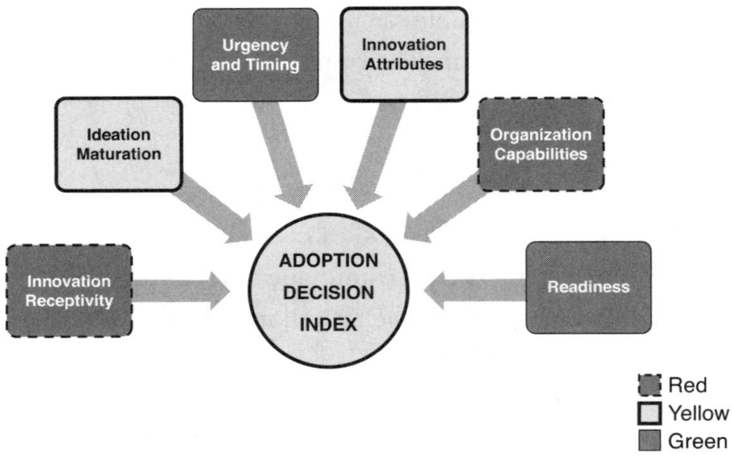

Figure 8.7 HICS adoption decision dashboard.

The dashboard results, including recommendations for bringing the innovation into the required green zone of the overall adoption index, will be presented to the executive management team.

Postscript

The CAO presented the proposal to the Executive Committee in January. The decision is still pending six months later. The initial reaction from the CEO was that the overall adoption decision index rating did not meet the threshold for consideration because it was in the yellow zone and the recommendations for improvement were not deemed sufficient. The proposal was tabled for two months.

The CAO followed up with the CEO in a one-on-one meeting to attempt to understand any underlying concerns. The biggest concern was that there were "just too many things going on." The CAO prodded to understand more. One of the concerns the CEO expressed involved how to implement the program through its new Medicaid managed care plan, and he said that it was too early to "rock the boat." He also confided that the medical director of the plan "had to go" in order to clear the way for physician communications and better partnership/conformity with management. He also emphasized that he was deep into making organization changes to flatten the organization and to change its character to be more responsive to employees and ideas. "This would take time."

Most staff at the VP level, as was the CAO, would move on to other priorities at this point given the obstacles expressed by the CEO. But the new CAO was young and not inured to the traditional insurance culture. She had a vision and commitment about demonstrating the value proposition of analytics based on its contribution to improving outcomes in the post fee-for-service era. She refused to think of her function as a cost center.

She "worked" the company network to build relationships and not necessarily to sell the innovation. She built trust among her colleagues and the Medicaid staff, including a number of physicians. She placed a priority on doing some analytics studies that the physicians had proposed. She had an open door and people saw her as someone who would listen to their concerns.

The environment changed. The health plan bought another large Medicaid managed care plan. However, the medical experience of the newly enrolled Medicaid patients was higher than expected and the Board questioned the viability of the strategy and issued a reprimand to the new CEO. The (problem) medical director of the Medicaid plan resigned, and one of the doctors that the CAO had befriended got the chief medical officer (CMO) job. The CMO had spent most of her career in community health centers and was keenly aware of the challenges in treating patients with multiple social and support challenges.

The CAO, acting as a matchmaker, introduced the CMO to the executive director of Health Leads. There was good chemistry and excitement about developing a pilot program. Health Leads supported the HICS analytics and wanted to incorporate them into their present assessments with patients; they liked the idea of adding other, external data to improve the identification of concerns for treatment success. The CAO also wanted to bring a new partner into the mix to demonstrate quick wins. Medication compliance was a promising area given that compliance was in the low 40% range. They had meetings with the marketing director and chief innovation officer of CQS, a retailer and pharmaceutical benefit manager.

The CAO scheduled a meeting with the CEO to touch base and inform him of the possibilities of this new partnership. The principals met with the CEO. He was impressed with how the design of the HICS was increasing the likelihood of success and reducing risk, and he scheduled another meeting of the executive committee two months out to consider it.

In the meantime, the CAO wanted to bring the adoption decision index to green. She figured that the organization capabilities factors would easily go to green given the plan to partner with Health Leads and CQS. She also had the support of the CMO and his willingness to provide leadership to physicians and care teams to accept the HICS. On the program model factor she felt that the previous results were sufficient but wanted to close down this potential "exit point." She worked with Health Leads to get funding from the foundation supporting it in exchange for the data on the performance of the program to help demonstrate its sustainability. This, in combination with CQS agreeing to provide funds for its aspect of the program and to share the prescription drug data, provided a good part of the operational funds needed to execute the plan.

She was skeptical about lifting the organization's innovation receptivity scores higher because culture change was a long-term investment and would take more time than she had. It was entwined with the reorg and she did not want to touch it. She was able to increase the overall score to yellow based on her continuing work with building relationships throughout the network.

She was also able to improve the ideation maturation score overall to green by demonstrating the linkages in cause and effect more clearly and by spending some time with those actuaries who were particularly skeptical. She convinced them that the project was derived from the banking industry and that she understood and respected their analytics challenges much more.

She was also able to address concerns about complexity and compatibility by clearly demonstrating how the program would work with partners and with analytics staff throughout the organization.

At the meeting, she would be able to demonstrate that the adoption index is in the acceptable green area. She knew there would be other concerns that might emerge and that she needed a well-thought-out position on them. She sifted information from the network and realized two other challenges. First was that the CEO was on the ropes about losses with the Medicaid programs. She worked with the actuaries and partners to "sharpen the pencil" on the ROI from HICS and was able to demonstrate a contribution to margin of $1 million in 12 months and an increase in the trend line on profitability for the Medicaid product line. She investigated the requirements for a CMS grant and demonstrated that the health plan had a good shot at receiving $500,000 if they demonstrated capability with outcomes measurement and improvement with underserved populations through HICS. The second thing she realized was that the CEO needed a public relations lift, and she positioned the program as "doing the right thing" for the community, which would help with marketing and his favorability ratings. Marketing was delighted with the key messages and with branding the plan as the only one in the market dedicated to the health and well-being of all of its members. The CEO received front-page coverage in the local papers as the champion of the new campaign.

The meeting with the CEO and his executive staff and principals from Health Leads, CQS, the foundation, the Medicaid medical director, the chief actuary, and the chief marketing officer lasted less than 30 minutes. The decision was unanimous to move forward with the adoption contingent upon the CAO writing the successful attainment of the Medicaid margin expectations into her performance review.

From start to finish the decision process took six months. The CAO took a well-deserved weekend vacation in the Bahamas, where she wrote the road map for her next analytics innovation. She was very concerned about the rising costs of diabetes and the analytics opportunity to identify obesity through the calculation of the BMI through external data. Her "back of the envelope" estimate of ROI exceeded that of the HICS. And she went about the task of addressing each of the 18 factors fueled by the glow of the Bahamian sunset and her success with moving the organization to adopt analytics innovations.

Endnotes

Chapter 1

1. E. McGlynn, S. Asch, J. Adams, et. al, "The Quality of Health Care Delivered to Adults in the United States," *New England Journal of Medicine* 348 (2003): 2635-2645.
2. "Analytics: The New Path to Value," a joint MIT Sloan Management Review and IBM Institute for Business Value study, Copyright © Massachusetts Institute of Technology 2010.
3. Steve Jobs, BrainyQuote.com, Xplore Inc, 2013, www.brainyquote.com/quotes/quotes/s/stevejobs416925.html, accessed February 25, 2013.
4. Paul Plsek, "Complexity and the Adoption of Innovation in Health Care," presentation to conference on Accelerating Quality Improvement in Health Care: Strategies to Speed the Diffusion of Evidence-Based Innovations, Washington, D.C., January 27, 2003.
5. de Bono, BrainyQuote.com, Xplore Inc, 2013, www.brainyquote.com/quotes/quotes/e/edwarddebo389925.html (accessed 3/29/13).
6. Charles Duhigg, "How Companies Learn Your Secrets," *New York Times*, February 16, 2012, www.nytimes.com/2012/02/19/magazine/shopping-habits.html?_r=1&ref=magazine.
7. Charles Duhigg, "Campaigns Mine Personal Lives to Get Out Vote," *New York Times*, October 13, 2012, www.nytimes.com/2012/10/14/us/politics/campaigns-mine-personal-lives-to-get-out-vote.html?pagewanted=all.

8. Bill Keller, "Invasion of the Data Snatchers," *New York Times*, January 13, 2013, www.nytimes.com/2013/01/14/opinion/keller-invasion-of-the-data-snatchers.html?pagewanted=all&_r=0.

9. James Manyika et al., "Big Data: The Next Frontier for Innovation, Competition, and Productivity," McKinsey Global Institute, May 2011, www.mckinsey.com/insights/mgi/research/technology_and_innovation/big_data_the_next_frontier_for_innovation.

10. William Eggers and John O'Leary, *If We Can Put a Man on the Moon: Getting Big Things Done in Government* (Boston: Harvard Business Press, 2009).

11. Everett Rogers, *Diffusion of Innovations*, 5th Edition (New York: Free Press, 2003).

12. Eggers and O'Leary, *If We Can Put a Man on the Moon.*

13. John Kingdon, *Agendas, Alternatives, and Public Policy* (Boston: Little Brown, 1984).

14. Y. Hasenfeld and T. Brock, "Implementation of social policy revisited: a political economy perspective," *Administration & Society* 22 (1991): 451-79.

15. J. Pressman and A. Wildavsky, *Implementation* (Berkeley, CA: University of California Press, 1984).

16. M. Gladwell, *The Tipping Point* (New York: Random House, 1999).

17. Paul Plsek, "Complexity and the Adoption of Innovation in Health Care," presentation to conference on Accelerating Quality Improvement in Health Care: Strategies to Speed the Diffusion of Evidence-Based Innovations, Washington, D.C., January 27, 2003.

18. R. D. Stacey, *Complex Responsive Processes in Organizations: Learning and Knowledge Creation* (New York: Routledge, 2001).

Chapter 2

1. S. A. McLeod, "Maslow's Hierarchy of Needs," 2007, retrieved from www.simplypsychology.org/maslow.html.

2. Preamble to the Constitution of the World Health Organization as adopted by the International Health Conference, New York, June 19-22, 1946, signed on July 22, 1946, by the representatives of 61 states (Official Records of the World Health Organization, no. 2, p. 100) and entered into force on April 7, 1948.

3. Micah Hartman et al., National Health Spending in 2011: Overall Growth Remains Low, but Some Payers and Services Show Signs of Acceleration, *Health Affairs* 32 (2013): 87-99.

4. Ibid.

5. Ibid.

6. Sarah Iselin, "Health Reform 2.0: From Coverage to Costs and the Role of Technology," April 24, 2012, www.slideshare.net/MassTLC/120424-masstlc-sarah-iselin-bcbsma-foundation-presentation.

7. Elizabeth H. Bradley and Lauren Taylor, "To Fix Health, Help the Poor," *New York Times*, www.nytimes.com/2011/12/09/opinion/to-fix-health-care-help-the-poor.html?_r=0.

8. E. Bradley et al., Health and Social Services Expenditures: Associations with Health Outcomes, *BMJ Quality & Safety* 20 (2011): 826-831.

9. OECD, *Health at a Glance 2011: OECD Indicators* (OECD Publishing, 2011), http://dx.doi.org/10.1787/health_glance-2011-en.

10. Ed. Steven H. Woolf and Laudan Aron, *U.S. Health in International Perspective: Shorter Lives, Poorer Health*, National Research Council and Institute of Medicine (2013), Panel on Understanding Cross-National Health Differences Among High-Income Countries.

11. Commonwealth Fund, "Why Not the Best? Results from the National Scorecard on U.S. Heath System Performance, 2011," www.commonwealthfund.org/Publications/Fund-Reports/2011/Oct/Why-Not-the-Best-2011.aspx?page=all.

12. Ibid.

13. E. McGlynn et al., "The Quality of Health Care Delivered to Adults in the United States," *New England Journal of Medicine* 348 (2003): 2635-45.

14. Ibid., 2635.

15. Steven A. Schroeder, "We Can Do Better—Improving the Health of the American People," *New England Journal of Medicine* 357 (2007):1221-28.

16. J. M. McGinnis, P. Williams-Russo, and J. R. Knickman, "The Case for More Active Policy Attention to Health Promotion," *Health Affairs* (Millwood) 21 (2002):78-93.

17. A. H. Mokdad et al., "Actual Causes of Death in the United States, 2000," *Journal of the American Medical Association* 291 (2004):1238-1245 [Erratum, *JAMA* 293 (2005):293-4, 298].

18. Schroeder, "We Can Do Better."

19. Alan Schwarz, "Attention Disorder or Not, Pills to Help in School," *New York Times*, October 9, 2012, www.nytimes.com/2012/10/09/health/attention-disorder-or-not-children-prescribed-pills-to-help-in-school.html?pagewanted=all&_r=0.

20. The Dartmouth Atlas of Healthcare, data by Hospital, www.dartmouthatlas.org/data/hospital/.

21. Atul Gawande, "The Cost Conundrum," *New Yorker Magazine*, June 1, 2009, www.newyorker.com/reporting/2009/06/01/090601fa_fact_gawande.

22. IBM Institute for Business Value, "The World's 4 Trillion Dollar Challenge," January 2010.

23. Don Berwick and Andrew Hackbarth, "Eliminating Waste in US Health Care," *JAMA* 307 (2012): 1513-16.

24. B. Temkin, "The Customer Experience Index, 2010," Forrester Research Inc., January 2010, https://docs.google.com/viewer?url=http%3A%2F%2Fwwwimages.adobe.com%2Fwww.adobe.com%2Fcontent%2Fdam%2FAdobe%2Fen%2Findustryinsights%2Fsolutions%2Fpdfs%2Fcustomer_experience_index_2010.pdf.

25. Kip Viscusi and Joseph E. Aldy, "The Value of a Statistical Life: A Critical Review of Market Estimates Throughout the World," *Journal of Risk and Uncertainty* 27 (2003): 5–76.

26. Berwick and Hackbarth, "Eliminating Waste in US Health Care."

27. Kaiser Family Foundation, "Family Health Premiums Rise 4 Percent to Average $15,745 in 2012, National Benchmark Employer Survey Finds," September 11, 2012, www.kff.org/insurance/ehbs091112nr.cfm.

28. Division of Health Care Financing and Quality, Commonwealth of Massachusetts, "Massachusetts Health Care Cost Trends: Price Variation in Health Care Services," May 2011, www.mass.gov/chia/docs/cost-trend-docs/cost-trends-docs-2011/price-variation-report.pdf.

29. Shubham Singhai, Jens Stueland, and Drew Ungerman, "How US Health Care Reform Will Affect Employee Benefits," McKinsey & Company, June 2011, http://www.mckinsey.com/insights/health_systems/how_us_health_care_reform_will_affect_employee_benefits.

30. Kaiser Permanente, "Fast Facts about Kaiser Permanente," accessed February 28, 2013, http://xnet.kp.org/newscenter/aboutkp/fastfacts.html.

31. Wikipedia, "Group Medical Practice in the United States," accessed February 28, 2013, http://en.wikipedia.org/wiki/Group_medical_practice_in_the_United_States.

32. American Hospital Association, "Fast Facts on US Hospitals," accessed February 28, 2013, www.aha.org/research/rc/stat-studies/fast-facts.shtml.

33. Stephen Swensen et al., "Cottage Industry to Postindustrial Care—The Revolution in Health Care Delivery," January 20, 2010, www.nejm.org/doi/full/10.1056/NEJMp0911199.

34. Commonwealth Fund, "Why Not the Best?"

35. CNNMoney, "Top Industries: Most Profitable 2009," accessed February 28, 2013, http://money.cnn.com/magazines/fortune/global500/2009/performers/industries/profits/.

36. Eric Lipton, "Fiscal Footnote: Big Senate Gift to Drug Maker," *New York Times*, January 19, 2013, www.nytimes.com/2013/01/20/us/medicare-pricing-delay-is-political-win-for-amgen-drug-maker.html?pagewanted=all&_r=0.

37. Jamoom E, Beatty P, Bercovitz A, et al. "Physician adoption of electronic health record systems: United States 2011." NCHS data brief, no 98. Hyattsville, MD: National Center for Health Statistics. 2012.

38. U.S. Department of Health & Human Services, "HHS Secretary Kathleen Sebelius Announces Major Progress in Doctors, Hospital Use of Health Information Technology," February 12, 2012, www.hhs.gov/news/press/2012pres/02/20120217a.html.

39. UPMC, "UPMC Fosters 'Personalized Medicine' with $100 Million Investment in Sophisticated Data Warehouse and Analytics," October 1, 2012, www.upmc.com/media/NewsReleases/2012/Pages/UPMC-Personalized-Medicine-Investment.aspx.
40. Pat Croskerry, "Clinical Decision Making and Diagnostic Error," presentation at Risky Business London, May 24, 2012, www.risky-business.com/talk-128-clinical-decision-making-and-diagnostic-error.html.
41. IBM, "Memorial Sloan-Kettering Cancer Center, IBM to Collaborate in Applying Watson Technology to Help Oncologists," March 22, 2012, www-03.ibm.com/press/us/en/pressrelease/37235.wss.
42. Larry Norton, "How Collaboration Between IBM And Memorial Sloan-Kettering Taps The Wisdom Of Physicians," *Huffington Post*, Blog, March 29, 2012. www.huffingtonpost.com/dr-larry-norton.
43. James Manyika et al., "Big Data: The Next Frontier for Innovation, Competition, and Productivity," McKinsey Global Institute, May 2011, www.mckinsey.com/insights/mgi/research/technology_and_innovation/big_data_the_next_frontier_for_innovation.
44. SAP, "Global Healthcare and Big Data," marketing brochure.
45. Basel Kayyali, et al. "The Big-Data Revolution in US Health Care: Accelerating Value and Innovation," April 2013, www.mckinsey.com/insights/health_systems/The_big-data_revolution_in_US_health_care.
46. American Diabetes Association, Diabetes Basics, "Diabetes Statistics," data from the 2011 National Diabetes Fact Sheet (released January 26, 2011), www.diabetes.org/diabetes-basics/diabetes-statistics/.
47. Jesse Gruman, "Hospitals, Practice Administrators and Clinicians: You Gotta Learn to Love Patient Ratings," Health Affairs Blog, http://healthaffairs.org/blog/2012/04/25/hospitals-practice-administrators-and-clinicians-you-gotta-learn-to-love-patient-ratings.

Chapter 3

1. U.S. Census, "Latest Annual Retail Trade Report, 2010," www.census.gov/retail/.

2. Plunkett Research, "Introduction to the Retail Industry," accessed February 28, 2013, www.plunkettresearch.com/retailing-stores-market-research/industry-and-business-data.

3. Kaiser Family statehealthfacts.org, "United States: Total Health Care Employment, 2011," accessed February 28, 2013, www.statehealthfacts.org/profileind.jsp?rgn=1&ind=445&cat=8.

4. Walmart 2012 Annual Report, "Five-Year Financial Summary," www.walmartstores.com/sites/annual-report/2012/WalMart_Financials.pdf.

5. CNNMoney, "Industries: Revenues, Profits, 2012," accessed February 28, 2013, http://money.cnn.com/magazines/fortune/fortune500/2012/industries/223/.

6. Nivedita Bhattacharjee and Jessica Wohl, "U.S. Retailers Scramble after Lackluster Holiday Sales," Reuters, December 26, 2012, www.reuters.com/article/2012/12/26/us-usa-retail-holiday-idUSBRE8BL03G20121226.

7. PricewaterhouseCoopers LLP and Kantar Retail LLC, "Retailing 2020: Winning in a Polarized World," September 2012, www.pwc.com/us/en/retail-consumer/publications/retailing-2020.jhtml.

8. National Retail Federation, "Profitability of Large Retail Corporations, 2006-2010," accessed February 28, 2013, www.nrf.com/modules.php?name=Pages&sp_id=1256#Profitability.

9. Index Mundi, "Commodity Prices," accessed February 28, 2013, www.indexmundi.com/commodities/.

10. Telecrunch, "Forrester Forecast: Online Retail Sales Will Grow to $250 Billion by 2014," March 8, 2010, http://techcrunch.com/2010/03/08/forrester-forecast-online-retail-sales-will-grow-to-250-billion-by-2014/.

11. WWWMetrics, "The Growth of Online Shopping," accessed February 28, 2013, www.wwwmetrics.com/shopping.htm.

12. Accenture, "Accenture 2010 Global Consumer Survey," February 2011, www.accenture.com/us-en/Pages/insight-accenture-customer-satisfaction-survey-2010-summary.aspx.

13. Customer Experience Matters, "Forrester's 2010 Customer Experience Rankings," January 11, 2010, accessed February 28, 2013, http://experiencematters.wordpress.com/2010/01/11/forrester%E2%80%99s-2010-customer-experience-rankings/.

14. Internet Retailer, "Top 500 Guide," accessed April 15, 2013, www.internetretailer.com/top500/profiles/BarnesandNoble-com-Inc/#BarnesandNoble-com-Inc.

15. National Retail Federation Stores, "2011 Top 100 Retailers," accessed April 15, 2013, www.stores.org/2011/Top-100-Retailers#.UN3mRuTAcuB.

16. Dhruv Grewal et al., *Retailing in the 21st Century: Current and Future Trends* (Berlin [u.a.]: Springer, 2005), 13-25.

17. IBM Institute for Business Value, "Customer Analytics Pay Off," January 2012, http://public.dhe.ibm.com/common/ssi/ecm/en/gbe03478usen/GBE03478USEN.PDF.

18. IBM Institute for Business Value, "From Stretched to Strengthened: Insights for the Global Chief Marketing Officer Study," 2011, http://public.dhe.ibm.com/common/ssi/ecm/en/gbe03433usen/GBE03433USEN.PDF.

19. Deloitte, "Smartphones Will Influence $689 Billion in Retail Store Sales by 2016: Deloitte Study," www.deloitte.com/view/en_US/us/press/Press-Releases/21b5a46ee8828310VgnVCM2000001b56f00aRCRD.htm.

20. Matt Anderson et al., "Turning 'Like' to 'Buy': Social Media Emerges as a Commerce Channel," booz&co, 2011.

21. Charles Duhigg, "How Companies Learn Your Secrets," *New York Times*, February 16, 2012, www.nytimes.com/2012/02/19/magazine/shopping-habits.html?_r=1&ref=magazine.

22. Marie T. Brown, MD, and Jennifer K. Bussell, MD, "Medication Adherence: WHO Cares?" *Mayo Clinic Proceedings* 86, no. 4 (2011): 304–314, www.ncbi.nlm.nih.gov/pmc/articles/PMC3068890/.

23. Fitbit Web site, www.fitbit.com/.

24. Custora Web site, https://www.custora.com/home/feature_customer_lifetime_value_clv.

25. Acxiom White Paper, Healthcare: Post Reform, 2011, www.acxiom.com/healthcare.

26. CVS Caremark 2011 Annual Report, http://drugstorenews.com/cvs-caremark-2011-annual-report.

27. T. Brennand, T. Dollear, and M. Hu, "An Integrated Pharmacy-Based Program Improved Medication Prescription and Adherence Rates in Diabetes Patients," *Health Affairs* 31 (2012):120-129.

28. J. Gruman, "I'm Not Taking that Drug If It Makes Me Itch! More on Medication Adherence" Prepared Patient Forum Blog, May 2, 2012, http://blog.preparedpatientforum.org/blog/2012/05/i%E2%80%99m-not-taking-that-drug-if-it-makes-me-itch-more-on-medication-adherence/.

29. Telenav, "Survey Finds One-Third of Americans More Willing to Give Up Sex Than Their Mobile Phones," August 3, 2011, www.telenav.com/about/pr-summer-travel/report-20110803.html.

30. E-Rewards Web site, www.e-rewards.com.

31. Health Prize Web site, https://www.healthprize.com/corporate/home.do;jsessionid=FDC9947F724EBE3E59CB09348A883C39.jvm43.

32. N. Singer, "Learning to Chase Online Word of Mouth," *New York Times*, May 26, 2012, www.nytimes.com/2012/05/27/business/fabcom-and-the-value-of-online-word-of-mouth.html?_r=1.

Chapter 4

1. Jeremy Ashkenas et al., "The 2012 Money Race: Compare the Candidates," *New York Times*, accessed August 21, 2012, http://elections.nytimes.com/2012/campaign-finance.

2. Editorial, "In Thrall to Sheldon Adelson," *New York Times*, August 16, 2012, www.nytimes.com/2012/08/17/opinion/in-thrall-to-sheldon-adelson.html.

3. Marc Thiessen, "Obama's 'Moneyball' Campaign," *Washington Post*, November 12, 2012, http://articles.washingtonpost.com/2012-11-12/opinions/35505532_1_obama-campaign-obama-strategist-obama-team.

4. Nate Silver, *The Signal and the Noise* (New York: Penguin Press, 2012).

5. Nate Silver, "Swing Voters and Elastic States," *New York Times*, May 21, 2012.

6. Ibid.
7. Rebecca Berg, "Few Voters Are Truly Up for Grabs, Research Suggests," *New York Times*, August 16, 2012.
8. OpenSecrets, "2012 Presidential Race," accessed October 14, 2012, www.opensecrets.org/pres12/index.php.
9. Dan Siroker, "How We Used Data to Win the Presidential Election," presentation at Google Partner Summit, 2009, Mountain View, CA, October 13, 2009, www.youtube.com/watch?v=xIpLwIRytMw.
10. "U.S. Ties Hospital Payments to Making Patients Happy," *Wall Street Journal*, October 14, 2012, http://online.wsj.com/article/SB10000872396390443890304578010264156073132.html.
11. National Center for Health Statistics, Average Length of Stay Report, www.cdc.gov/nchs/data/nhds/1general/2009gen1_agesexalos.pdf.
12. American Hospital Association, "Fast Facts on US Hospitals," accessed April 15, 2013, www.aha.org/research/rc/stat-studies/fast-facts.shtml.
13. Canadian Index of Well-Being, "How Are Canadians Really Doing?" The 2012 CIW Report (Waterloo, ON: Canadian Index of Well-Being and University of Waterloo, 2012), https://uwaterloo.ca/canadian-index-wellbeing/sites/ca.canadian-index-wellbeing/files/uploads/files/CIW2012-HowAreCanadiansReallyDoing-23Oct2012_0.pdf.
14. County Health Rankings and Roadmaps Web site, www.countyhealthrankings.org/.
15. CDC, Behavioral Risk Factor Surveillance System Web site, www.cdc.gov/brfss/.
16. Massachusetts Department of Public Health, "A Profile of Health Among Massachusetts Adults in Selected Cities, 2008," Results from the Behavioral Risk Factor Surveillance System, October 2009, www.mass.gov/eohhs/docs/dph/behavioral-risk/cities-08.pdf.
17. FiveThirtyEight, "Methodology," *New York Times*, accessed February 28, 2013, http://fivethirtyeight.blogs.nytimes.com/methodology/.
18. TargetPoint Web site, www.targetpointconsulting.com/.
19. Constituent Cloud Web site, www.constituentcloud.com/.
20. Ibid.

21. Eimar Boesjes, "Big Data in Politics: Mobile, Instant, Ubiquitous," Strata Conference, Get the (Data) Vote Out, June 20, 2012, http://oreillynet.com/pub/e/2311.

22. NationBuilder Web site, http://nationbuilder.com/.

23. Jim Rutenberg and Jeff Zeleny, "Obama Mines for Voters with High-Tech Tools," *New York Times*, March 8, 2012, www.nytimes.com/2012/03/08/us/politics/obama-campaigns-vast-effort-to-re-enlist-08-supporters.html?pagewanted=all.

24. Strata Conference, Get the (Data) Vote Out.

25. Acxiom Web site, www.acxiom.com/.

26. Experian Web site, www.experian.com/.

27. Equifax Web site, www.equifax.com/home/en_us.

28. Spokeo Web site, www.spokeo.com/.

29. Charles Duhigg, "Campaigns Mine Personal Lives to Get Out the Vote," *New York Times*, October 13, 2012, www.nytimes.com/2012/10/14/us/politics/campaigns-mine-personal-lives-to-get-out-vote.html?pagewanted=all.

30. Ibid.

31. Ibid.

32. Ethan Roeder, "I Am Not Big Brother," *New York Times*, December 5, 2012, www.nytimes.com/2012/12/06/opinion/i-am-not-big-brother.html?hp.

33. Strata Conference, Get the (Data) Vote Out.

34. Ibid.

35. http://usatoday30.usatoday.com/news/politics/twitter-election-meter, accessed April 15, 2013.

Chapter 5

1. Liz Rappaport and Dan Fitzpatrick, "Pain Spreads to Biggest Banks," *Wall Street Journal*, October 19, 2011, http://online.wsj.com/article/SB10001424052970204479504576638653920110530.html.

2. McKinsey & Company, "In Search of a Sustainable Model for Global Banking," *McKinsey Quarterly*, September 2011, www.mckinsey.com/features/global_banking.
3. Plunkett Research, "Introduction to the Banking, Mortgages, and Credit Industry," www.plunkettresearch.com/banking-mortgages-market-research/industry-and-business-data.
4. B. Temkin, "The Customer Experience Index, 2010," Forrester Research Inc., January 2010, https://docs.google.com/viewer?url=http%3A%2F%2Fwwwimages.adobe.com%2Fwww.adobe.com%2Fcontent%2Fdam%2FAdobe%2Fen%2Findustryinsights%2Fsolutions%2Fpdfs%2Fcustomer_experience_index_2010.pdf.
5. IBM Center for Applied Insights, "The Value of Building Sustainable Health Systems," January 2012, http://public.dhe.ibm.com/common/ssi/ecm/en/hhe12345usen/HHE12345USEN.PDF.
6. Ibid.
7. Abhishek Mehta (moderator), Big Data in Banking (panel presentation), Strata Summit: The Business of Data, New York, NY, September 20, 2011, www.youtube.com/watch?v=DfmEMrkt8BY.
8. McKinsey & Company, "The Triple Transformation: Achieving a Sustainable Business Model," 2nd McKinsey Annual Review on the Banking Industry, October 2012.
9. Plunkett Research, "The Banking, Mortgages, and Credit Industry."
10. Jed Horowitz, "McKinsey Casts Gloomy Eye on World Banking," Reuters, October 8, 2012, www.reuters.com/article/2012/10/09/us-mckinseystudy-banks-idUSBRE89800020121009.
11. www.paynearme.com/.
12. https://www.greendot.com/greendot.
13. https://www.netspend.com/.
14. Andrew Jennings, "Stealing Some Magic from Retailers," http://bankinganalyticsblog.fico.com/2012/11/stealing-some-magic-from-retailers.html.
15. www.prosper.com/.

16. Julie Wooding, "When the Game Changes, Adapt the Fundamentals," presentation to FICO World 2011, November 3, 2011.

17. Plunkett Research, "The Banking, Mortgages, and Credit Industry."

18. McKinsey & Company, "The Triple Transformation."

19. Ibid., 32.

20. Ibid., 23.

21. "Big Data: Crunching the Numbers," *Economist*, May 19, 2012, www.economist.com/node/21554743.

22. Jeanne Capachin and Randy Perry, "Banking for Success: Using Analytics to Grow Wallet Share," IDC White Paper, January 2012.

23. Accenture, "Banking 2016: Accelerating Growth and Optimizing Costs in Distribution and Marketing," 2012, www.accenture.com/SiteCollectionDocuments/PDF/FinancialServices/Accenture-Banking-2016.pdf.

24. Steven Pilloff and Anthony Santomero, "The Value Effects of Bank Mergers and Acquisitions, 1997."

25. McKinsey, "The Triple Transformation," 32.

26. Michael Simkovic, "Secret Liens and the Financial Crisis of 2008," *American Bankruptcy Law Journal* 83 (2009): 253.

27. Jihad Dagher and Ning Fu, "What Fuels the Boom Drives the Bust: Regulation and the Mortgage Crisis," IMF Working Paper WP/11/215 (2011), www.imf.org/external/pubs/ft/wp/2011/wp11215.pdf.

28. McKinsey & Company, "Banking on Customer Centricity," EMEA Banking Practice, April 2012, www.mckinsey.com/clientservice/Financial_Services/Capabilities/~/media/Reports/Financial_Services/Customer_Centricity.ashx.

29. https://www.wellsfargo.com/invest_relations/vision_values/3.

30. Chisoo Lyons, "CFO Focus: Deepen Member Connections," *CU Management* 35, no. 6 (2012), www.cumanagement.org/article/view/id/Cfo-focus-deepen-member-connections.

31. "Crunching the Numbers."

32. Ibid.

33. Sean Rowles, "Retail Banking Growth and the Critical Role of Risk," presentation to FICO World 2011, November 2, 2011.
34. NationMaster, Health Statistics, www.nationmaster.com/graph/hea_phy_per_1000_peo-physicians-per-1-000-people.
35. Bonnie Kavoussi, "IBM 'Watson' Computer to Work at Citigroup," *Huffington Post*, March 6, 2012, www.huffingtonpost.com/2012/03/06/ibm-watson-citigroup_n_1323467.html.
36. Tom Groenfeldt, "IBM's Watson, Cedars-Sinai and WellPoint Take On Cancer," *Forbes*, February 1, 2012, www.forbes.com/sites/tomgroenfeldt/2012/02/01/ibms-watson-cedars-sinai-and-wellpoint-take-on-cancer/3/.
37. Capachin and Perry, "Using Analytics."
38. IBM Center for Applied Insights, "The Value of Building Sustainable Health Systems."
39. Penny Crosman, "The Big Data Myth," *American Banker*, June 1, 2012, www.americanbanker.com/btn/25_6/customer-data-improve-marketing-1049578-1.html.
40. David Paul Morris, "Health-care Apps for Smartphones Pit FDA Against Tech Industry," *Washington Post*, June 22, 2012, www.washingtonpost.com/business/economy/health-care-apps-for-smartphones-pit-fda-against-tech-industry/2012/06/22/gJQAHCcBvV_story_1.html.
41. Matt Richtel, "Apps Alert the Doctor When Trouble Looms," *New York Times*, October 8, 2012, http://well.blogs.nytimes.com/2012/10/08/apps-alert-the-doctor-when-trouble-looms/.
42. https://www.skinofmine.com/.
43. Jan Hoffman, "Texting the Teenage Patient," *New York Times*, October 8, 2012, http://well.blogs.nytimes.com/2012/10/08/texting-the-teenage-patient/.
44. FICO, "Understanding Predictive Analytics," www.fico.com/en/Communities/PredictiveAnalytics/Pages/default.aspx.
45. Gordon Cameron, "Revitalizing Growth in the Reset Economy," presentation to FICO World 2011, November 1, 2011, http://bankinganalyticsblog.fico.com/2011/11/pncs-gordon-cameron-calls-for-a-new-approach-to-lending.html.

46. Penny Crosman, "The Big Data Myth."
47. www.zestcash.com/welcome.
48. https://www.wonga.com/.
49. www.cignifi.com/en-us/.
50. www.tesco.com/.
51. "Crunching the Numbers."
52. http://www.corelogic.com/industry/consumer-finance-solutions.aspx.
53. Larry Rosenberger, "Bringing Customers to the Forefront of Analytics," presentation at FICO World 2011, November 2, 2011.
54. Capachin and Perry, "Using Analytics."
55. www.healthleadsusa.org/our-model/.

Chapter 6

1. Wikipedia, "Top 10 Leagues in Total Attendance," http://en.wikipedia.org/wiki/List_of_sports_attendance_figures#Top_10_Leagues_in_total_attendance.
2. Wikipedia, "World Series Television Ratings," http://en.wikipedia.org/wiki/World_Series_television_ratings.
3. Wikipedia, "Top 10 Leagues in Total Attendance."
4. Plunkett Research, "Introduction to the Sports Industry," www.plunkettresearch.com/sports-recreation-leisure-market-research/industry-and-business-data/statistics.
5. Ibid.
6. Suzanne Sataline and Shirley Wang, "Medical Schools Can't Keep Up," *Wall Street Journal*, April 12, 2010, http://online.wsj.com/article/SB10001424052702304506904575180331528424238.html.
7. American Gaming Association, "Sports Wagering," www.americangaming.org/industry-resources/research/fact-sheets/sports-wagering.

8. Doug Henschen, "Analytics Drives Next Generation of Moneyball in Sports," *Information Week*, March 8, 2012, report on the MIT sports forum, www.informationweek.com/software/business-intelligence/analytics-drives-next-generation-of-mone/232602229.

9. David Holmes, "Who (Or What's) Best at Predicting March Madness Winners?" Fast Company, March 12, 2012, www.fastcompany.com/1824382/who-or-whats-best-predicting-march-madness-winners.

10. The Richest, "Highest-Paid Athletes—World's Richest Athletes," June 19, 2012, www.therichest.org/sports/forbes-highest-paid-athletes/.

11. Ibid.

12. Michael Lewis, *Moneyball: The Art of Winning an Unfair Game* (New York: W.W. Norton & Company, 2003), 15.

13. Ibid., 59.

14. Ibid., 249.

15. Chuck Salter, "In Relentless Jock-Nerds War, Hope for Peace Through Analytics," Fast Company, March 14, 2012, www.fastcompany.com/1824499/relentless-jocks-nerds-war-hope-peace-through-analytics.

16. Nate Silver, *The Signal and the Noise: Why So Many Predictions Fail—but Some Don't* (New York: Penguin Press, 2012), 95.

17. *Moneyball*, 61.

18. NBA, "NBA Player Field Goal Shooting Statistics, 2012-13," http://espn.go.com/nba/statistics/player/_/stat/field-goals/sort/fieldGoalPct.

19. Kirk Goldsberry, "CourtVision: New Visual and Spatial Analytics for the NBA," paper submitted to MIT Sloan Sports Analytics Conference 2012.

20. Ira Boudway, "Baseball: Running the New Numbers," *Bloomberg Businessweek*, March 31, 2011, www.businessweek.com/magazine/content/11_15/b4223072802462.htm.

21. Consumer Reports.org, "How to Choose a Doctor," May 2008, www.consumerreports.org/cro/2012/12/how-to-choose-a-doctor/index.htm.

22. Massachusetts Health and Human Services, Board of Registration in Medicine, www.mass.gov/eohhs/gov/departments/borim/.

23. Massachusetts Health and Human Services, MyHealthCare Options, http://hcqcc.hcf.state.ma.us/.
24. *Moneyball*, 76.
25. Nate Silver, *The Signal and the Noise* (New York: The Penguin Press, 2012), 85.
26. Ibid., 90.
27. Robert Ayer, "Big 2's and Big 3's: Analyzing How a Team's Best Players Complement Each Other," presentation at MIT Sloan Sports Analytics Conference, March 2, 2012, www.sloansportsconference.com/wp-content/uploads/2012/02/Robert-Ayer-Sloan-Sports-Analytics-Conference.pdf.
28. Gartheeban Ganeshapillai and John Guttag, "Predicting the Next Pitch," presentation at MIT Sloan Sports Analytics Conference, March 2, 2012, www.sloansportsconference.com/wp-content/uploads/2012/02/98-Predicting-the-Next-Pitch_updated.pdf.
29. Simon Williams and Jacomo Corbo, "Data Science for Agile Strategy: From Formula 1 to the Boardroom," presentation to Strata Conference on Making Data Work, October 1, 2012, http://strataconf.com/strataeu/public/schedule/detail/26030.
30. Rajiv Maheswaran et al., "Deconstructing the Rebound with Optical Tracking Data," presentation at MIT Sloan Sports Analytics Conference, March 2, 2012, www.sloansportsconference.com/wp-content/uploads/2012/02/108-sloan-sports-2012-maheswaran-chang_updated.pdf.
31. FanGraphs, "What Is PITCHF/x," www.fangraphs.com/library/index.php/misc/pitch-fx/.
32. Boudway, "Baseball: Running the New Numbers."
33. Steve Ham, "Rugby: How Analytics Can Reduce Injuries in a Very Tough Sport," *Huffington Post*, May 1, 2012, www.huffingtonpost.com/steve-hamm/rugby-how-analytics-can-r_b_1467705.html.

Chapter 7

1. A note on possible bias: The adaptations discovery process examined industry context, strengths, and associated analytic sweet spots. The process looked for emerging insights and was not driven by any hypotheses related to healthcare. However, as discussed in Chapter 8, our observations and beliefs are shaped by our confirmation bias; that is, we seek out information that confirms our notions about how the world works and avoid information that does not. The author's belief in the importance of activating people in healthcare might have influenced the interpretation of how the sweet spots are translated for healthcare.

Chapter 8

1. J. Pressman and A. Wildavsky, *Implementation* (Berkeley, CA: University of California Press, 1984).
2. Everett Rogers, *Diffusion of Innovations*, 5th Edition (New York: Free Press, 2003).
3. William Eggers and John O'Leary, *If We Can Put a Man on the Moon... Getting Big Things Done in Government* (Boston: Harvard Business Press, 2009).
4. John Kingdon, *Agendas, Alternatives, and Public Policy* (Boston: Little Brown, 1984).
5. Y. Hasenfeld and T. Brock, "Implementation of Social Policy Revisited: A Political Economy Perspective," *Administration and Society* 22 (1991): 451–79.
6. Pressman and Wildavsky, *Implementation*.
7. M. Gladwell, *The Tipping Point: How Little Things Can Make a Big Difference* (New York: Little Brown and Company, 2000).
8. Paul Plsek, "Complexity and the Adoption of Innovation in Health Care," presentation to conference on Accelerating Quality Improvement in Health Care: Strategies to Speed the Diffusion of Evidence-Based Innovations, Washington, D.C., January 27, 2003.

9. A. Pettigrew, E. Ferlie, and L. McKee, *Shaping Strategic Change: Making Change in Large Organizations: The Case of the National Health Service* (London: Sage, 1992).
10. Don Berwick, "Disseminating Innovations in Health Care," *Journal of the American Medical Association* 289, no. 15 (2003):1969-75.
11. http://en.wikipedia.org/wiki/Primum_non_nocere.
12. Eggers and O'Leary, *If We Can Put a Man on the Moon*.
13. Hasenfeld and Brock, "Implementation of Social Policy."
14. Berwick, "Disseminating Innovations in Health Care."
15. Paul Plsek, "Complexity and the Adoption of Innovation in Health Care," presentation to conference on Accelerating Quality Improvement in Health Care Strategies to Speed the Diffusion of Evidence-Based Innovations, Washington, D.C., January 27, 2003.
16. Gladwell, *The Tipping Point*.
17. Judith Hibbard, Jessica Greene, and Valerie Overton, "Patients with Lower Activation Associated with Higher Costs; Delivery Systems Should Know Their Patients' 'Scores,'" *Health Affairs* 32, no. 2 (2013): 216-222.

Index

A

D

I

Q-R

S

T

V

W-Z